combinatorics
graphs and algebra

methods and models
in the social sciences

5

MOUTON · THE HAGUE · PARIS

combinatorics
graphs and algebra

by CENTRE DE MATHEMATIQUE SOCIALE
ÉCOLE DES HAUTES ÉTUDES EN SCIENCES SOCIALES

MOUTON · THE HAGUE · PARIS

This volume has been published with the support
of the Department of Social Sciences of Unesco, Paris

ISBN: 90 279 7511 6

Translated from the French by: Victoria von Hagen

Cover design by Jurriaan Schrofer

Printed in the Netherlands

preface

The purpose of this manual, is to introduce students and researchers in the human and social sciences to some mathematical domains of particular importance because of their potential application in these sciences.

This first volume is devoted to fundamental chapters in algebra and combinatorics. It has been conceived as an introduction to the meaning and proper usage of certain key words which are encountered more and more frequently in the construction of models or in the presentation of mathematized theories in psychology, sociology, anthropology, linguistics or musical composition. The *index of terms* found at the end of this manual is thus of major importance: for each term, this index refers the reader to a definition and to the chapters in this text where the term has been used in various contexts.

These contexts . . . what are they? Those found in the most *elementary* 'structures' which can be assigned a finite set or a set which can be finitely constructed, structures of ordering, classifications, trees, Boolean algebras, monoids, groups, simplexes and measure scales. Moreover, these topics are now part of what is commonly taught in many universities, especially in French higher education, to students in the social sciences.

The various articles in this manual have been written so that they can be read independently: that is, the order in which they can be read is of little importance and is left to the discretion of the reader. Nevertheless, we have made every effort to show the interrelationships that unite the domains introduced in this volume by referring the reader to various other articles. In addition, the bibliography of each chapter furnishes references which will permit the reader to improve on this brief introduction to these mathematical domains, as well as acquainting him with the detailed applications of these topics in the human sciences.

The reader is not supposed to have previous mathematical knowledge, other than that normally acquired in high school. However, those readers who encounter too much difficulty can refer to introductory texts such as those already published by certain authors of this manual[1]. In these works the reader will find the elementary bases of combinatorics (subsets of a finite set and their simplicial organization) to which the articles here sometimes refer. The reader can also consult the *symbol index* placed at the beginning of this manual.

M. Barbut

1. M. Barbut, C. d'Adhémar, B. Leclerc and P. Jullien, *Mathématiques élémentaires, applications à la statistique et aux sciences sociales,* Paris, P.U.F., 1973; M. Barbut, *Mathématiques des sciences de l'homme,* Paris, P.U.F., 1968; G. Th. Guilbaud, *Mathématiques,* Paris, P.U.F., 1966; P. Rosenstiehl and J. Mothes, *Mathématiques de l'action,* Paris, Dunod, 1965.

contents

3. Boolean algebras, Boolean rings
by G. Th. Guilbaud

4. Simplicial objects
by G. Th. Guilbaud

5. Monoids and groups
by C. d'Adhémar

6. Measure scales
by M. Barbut

symbol index

Symbols	Use and meaning	Commentary (remarks)
{ }	$E = \{2, \square, b, \star\}$: the set E and its element	the set E and the list of its elements: 2, $\square$, b, $\star$
$\lvert E \rvert$, card. E	$\lvert E \rvert = 4$: the cardinal of E is four	Number of elements in E
$\in$	$2 \in E$: 2 is an element of / or belongs to the set E	
(. . . , . . .)	(x, y): the ordered pair xy	starting with a set E we can construct a new unique product set which is the set of pairs (x, y) such that $x \in E$ and $y \in E$. The notion of ordered pair has a privileged meaning in set theory, but the number of components can be increased: triplet, quadruplet, . . . n-uplet
$E \times E = E^2$	E cross E, or E to the second power	the cartesian product of the set E by itself is the set of pairs (x, y) where $x \in E$ and $y \in E$
$\subset$	$X \subset Y$: the set X is included in the set Y	$X \subset Y$ if, and only if, for every x we have $x \in X \Rightarrow x \in Y$

Symbols	Use and meaning	Commentary (remarks)
$\{x\}$	the set reduced to a single element x	
$=$	$X = Y$: the set X is equal to the set Y	the equality of the two sets is not satisfied unless they contain the same elements
$\forall$	$(\forall x \in X)\ (x = x)$: for every $x \in X$ we have $x = x$	the proper use of these two symbols, called logical quantifiers requires a formalism that exceeds the scope of this work; they are only used here as abbreviations in the context indicated here
$\exists$	$(\exists x \in X)\ (x = y)$: there exists *some* $x \in X$ such that we have $x = y$	
$\Rightarrow$	$x \Rightarrow y$: x implies y	if x, then y
$\Leftrightarrow$	$x \Leftrightarrow y$: x if, and only, if, y	
$\cup$	$X \cup Y$: X union Y	$X \cup Y$ designates the set formed by both the elements of X and the elements of Y (and of X and Y simultaneously)
P^C, CP, $\bar{P}$, $E-P$, $\complement_E P$	the subset of the set E that is the compliment of the subset P of E	$P \cup P^C = E$; the union of P and its complementary subset reconstitutes the set E
$\cap$	$X \cap Y$: X intersection Y	$X \cap Y$ designates the set formed by the elements that belong to both X and Y
$\mathcal{P}(E)$ 2^E	the set of the subsets of the set E	if $\lvert E \rvert = 4, \lvert \mathcal{P}(E) \rvert = 2^4$. Two special subsets are E itself (the whole set) and the empty subset, or ϕ

Symbols	Use and meaning	Commentary (remarks)
ϕ	$\phi \in \mathcal{P}(E)$: the empty set is an element of the set of the subsets of E	the empty set is the set which contains no elements. It is by definition an element of the set of the subsets of the set under consideration, E
$x R y$ or $(x, y) \in R$	xRy: the pair (x, y) belongs to or satisfies the (binary) relation R	
$x \not R y$	(x, y) does not satisfy the relation R	
$x \neq y$	x is different from y	
$x \equiv y$ $x \sim y$ $x \equiv y \pmod R$	x is equivalent to y x is equivalent to y, or x is equivalent to y modulo R	 x is equivalent to y from the point of view of the relation R
$\bigcup$	$\bigcup_{i \in I} X_i$: union of the family of the sets X_i	the union applies to a family of sets indexed by I
$E/R = \Pi_R$	the quotient set E modulo R = the partition induced by the equivalence relation R	the set of equivalence classes of the equivalence relation R is the same thing as the partition of E associated with the relation R
$f: E \to F$	the mapping f of the set E into the set F	
$x \to y$ or $x \to y$	y is the value (image) of x by the mapping (function) under consideration	
$f(x) = y$	f of x equals y	the image of x by the mapping f is the element y

Symbols	Use and meaning	Commentary (remarks)
$<$ and $\leqslant$	$x<y$: x is strictly smaller than y $x \leqslant y$: x is less than or equal to y	
$<$	$x<y$: x is covered by y or y covers x	$x<y$ and there is no element z such that $x<z<y$
$\vee$	$x \vee y$, x supremum y: the least upper bound of x and y	this element is the smallest of the upper bounds common to x and y
$\vee A$ or sup A	supremum of A	the smallest of the upper bounds common to all the elements of A
$\wedge$	$x \wedge y$: infimum of x and y	this element is the greatest of the lower bounds of x and y
$\wedge A$ or inf A	infimum of A	the greatest of the lower bounds common to all the elements of A
$[x, y]$	closed interval of xy	it is the set of the elements z such that $x \leqslant z \leqslant y$
$]x, y]$	the interval xy, open to the left and closed on the right	it is the set of z such that $x<z \leqslant y$
$\bar{R}$	transitive closure of the relation R	
$\boldsymbol{\wedge}$ or 0 $\boldsymbol{\vee}$ or 1	the poles or distinct elements of a Boolean algebra	$A \wedge A^c = \boldsymbol{\wedge}$ $A \vee A^c = \boldsymbol{\vee}$
$\oplus$	$A_1 \oplus A_2$: the direct sum of the abelians A_1 and A_2	the addition is defined on the set of the pairs by the rule: if $x = (x_1, x_2)$, $y = (y_1, y_2)$, then $x + y = (x_1 + y_1, x_2 + y_2)$
!	$n!$: n factorial	it is the product of the first n integers

Symbols	Use and meaning	Commentary (remarks)
N	the integers	$= \{0, 1, 2, 3, \ldots, n, \ldots\}$
Z	the relative integers	$= \{\ldots, -n, \ldots, -2, -1, 0, 1, 2, \ldots, n, \ldots\}$
Q	the rational numbers	$= \{r \mid r = \frac{p}{q}; p, q, \in \mathbf{Z} \text{ and } q \neq 0\}$
R	the real numbers	
s	$s(0)$: successor of zero	
$h = fog$	h is equal to f circle g	this is the compostion of mappings: the mapping g followed by the mapping f. fog is thus only defined if the initial set of f is identical to the final set of g

1. trees

by P. Rosenstiehl *and* B. Leclerc

Introduction

A systematic study of trees is certainly an elementary subject. After the straight line, what can be studied? The tree.

It is surprising that this subject still fails to occupy a proper place in mathematical text books. Yet algebra and combinatorics, as well as the applications of these subjects to the human sciences, are rife with trees. Moreover, trees are an ideal mathematical object to illustrate the interconnections of combinatorics with algebra.

Primarily, trees are a combinatorial subject. Everbody knows how to join n points with $n-1$ lines to form a tree; or to glue $n-1$ sticks end to end, creating a new node with each stick, giving thus n nodes. Such relations are spoken of in terms of injections.

Soon algebraic subjects make their appearance. *The free groups* include the paths in a tree; the retracings back and forth. In coding, it is linguistics that quickly dominate the subject. The tree is translated into words and becomes again a line; this will be discussed in section 1.

A tree can be rooted at any one of its points. It is organized then from this root point in a partial order: *semi-lattice* rooted trees. The evoluative approach, the hierarchical organization of the possibilities and the recurrence in the choices are the subject of section 2.

Finally the trees form a skeleton for all networks, and most calculations on networks are based on that minimal part. Anything that can be formalized as a network or a graph, a sociogram or a flow chart is based, to a certain extent, on an understanding of the rooted tree skeleton. Boundary algebra *the linear calculus* of cycles, cocycles, flows and tensions are presented from this point of view in section 3.

In our chapter on trees we will give eleven definitions of trees, all dissimilar, some even surprising, although all logically equivalent. Sometimes the object is a word, sometimes a set, a mapping or a graph. Each new definition introduced is in some way related to the preceding ones. Our reasoning can be presented schematically by a tree, naturally. We have given some complimentary exercise material and, even more important, provided suggestions for further reading in the 25 references cited in the final bibliography, pp. 55–57.

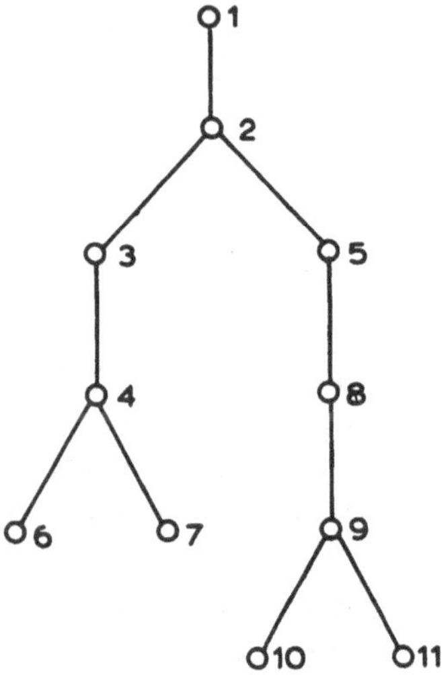

Fig. 1. *Successive definitions of trees*

1. Tree words

How can a tree be expressed? This is the first question we shall consider. To express a tree is first to define it and also put it into a communicable form so that certain information contained in it can be analyzed. Thus, in the first section we will define a tree and its coding.

Let us begin with a game. We put four points on a sheet of paper and try to find all the possible ways to join these points with lines to obtain a tree. It is immediately apparent that three lines are always required. With a little patience we can find the 16 trees obtained by joining the four given points in every possible way.

The herborium is shown in Figure 2 and Figure 3 below. The first is a graphic representation, the second its coding. How were these trees obtained? We chose four letters *a, b, c* and *d* (just as we chose four points); these symbols, it should be noted, have an alphabetic order. With these letters we shall write words composed of two letters, and the 16 words of length two thus obtained is the herborium – each word designating a tree. The bijective correspondence[1] is not immediately obvious, but will be established in the following paragraphs, thereby giving the reader a better idea of what a tree is. In the second part of this section another code will reveal still other properties of trees.

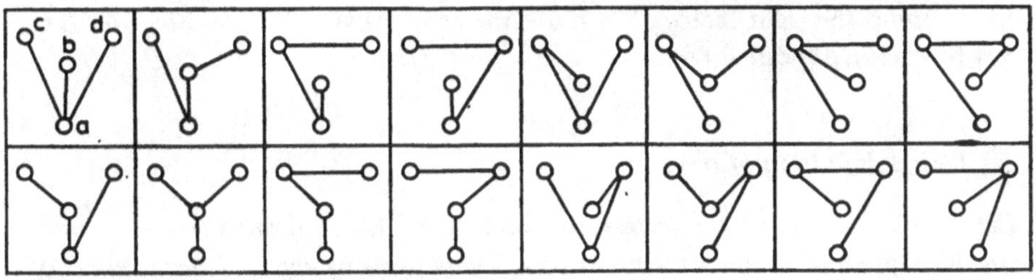

Fig. 2. *The 16 possible trees generated by joining 4 points* a, b, c, d

aa	ab	ac	ad	ca	cb	cc	cd
ba	bb	bc	bd	da	db	dc	dd

Fig. 3. *Coding of the 16 possible trees generated by joining 4 points* a, b, c, d

1.1 Free word trees

Henceforth, any word can designate a tree. Let us first define what we mean by a word. Given a finite set A called the alphabet:

$A = \{d, a, b, c\}$ for example,

any sequence of the elements of A, such as

badadaa

is called a word written in the alphabet A.

To concatenate a first word with a second means to place the second word

1. Between two sets A and B having the same cardinality, a bijective correspondence (or bijection) is a correspondence that associates to each element of A a unique element of B so that each element of B is associated with a unique element of A.

after the first. By definition, the concatenation of two words is another word.

badadaa is a word having seven occurrences of letters, with two occurrences of the letter *d* in the third and fifth location. It is also the concatenation of *bada* and *daa*. Single letters are words with a one letter occurrence. A word having *k* occurences is the concatenation (a noncommutative operation) of *k* words of one occurrence.

If *x* is the concatenation of the two words α and β ($\sigma = \alpha\beta$), α is called the left factor and β the right factor of σ. If σ is the concatenation of the three words α, β, γ ($\sigma = \alpha\beta\gamma$), β is called factor of σ.

1.1.1 First definition of trees

Let A *be an alphabet of* n *letters, totally ordered. We shall call a tree – or more specifically, a tree joining the points of* A *– any word having* n–2 *occurrences of the letters of* A.

As an example, let us consider the ordered alphabet:

A: abcd

the words:

aa ba da cc

are, by virtue of the above definition, trees joining the four points *a, b, c* and *d*. These words have a given numbers of occurrences but are otherwise unrestricted, without grammar; they are sometimes called free.

How do we draw the tree *da*? Since the letter *a* is distinguished by the alphabetic order, we can agree that the desired tree is rooted at *a*. The first operation is to replace *'da'* by *'ada'* (put *a* first).

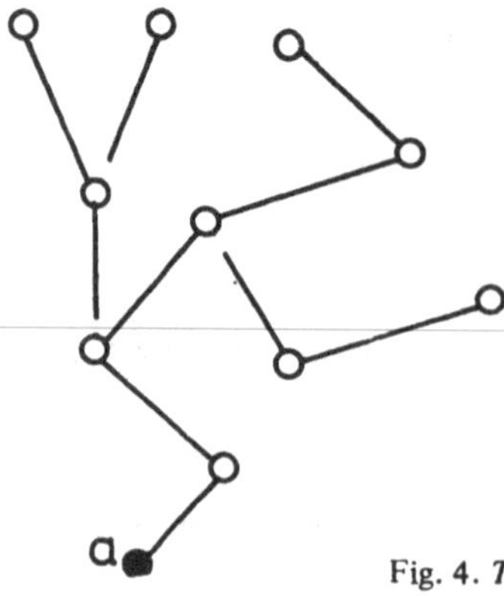

Fig. 4. *The principle of coding a tree by grafting its four branches*

The tree is to be seen as a branch having grown from *a*, then another branch grafted onto the first, the next on to the preceding, etc. These branches are words which are grafted by their first letter which has already appeared in the preceding branch. It should be noted that each branch is a subword of the word under consideration but lengthened by one letter which is the first of the absent letters of this word not yet grafted.

Let us trace '*da*' written in the alphabet *abcd.* In *ada, b* and *c* are absent.

The branches are: *adb*
ac

and we graft:

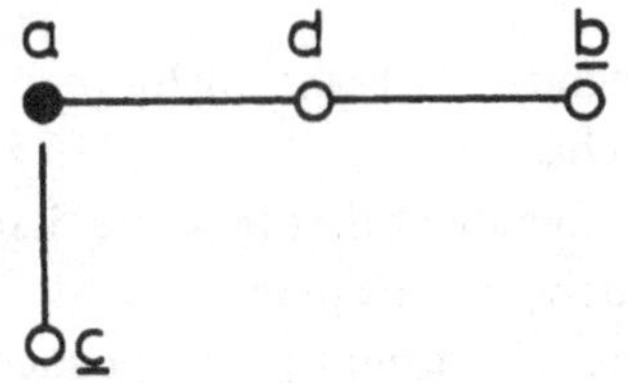

Fig. 5

Let us consider another example: *bafaf* written in the alphabet *abcdefg.* In *abafaf, cdeg* are absent.

The branches are: *abc*
afd
ae
fg

and we graft:

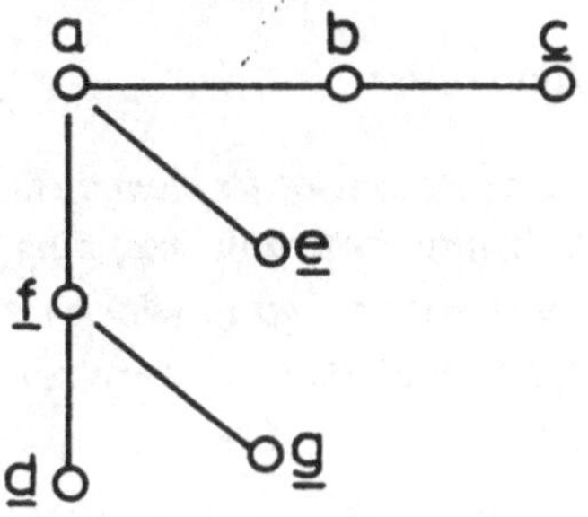

Fig. 6

a, the leading letter which has been written first is called the root. Any letter of the word already written in the word designates a graft point. That is, the beginning of a new branch.

1.1.2 Algorithm for tracing a tree

Let σ be a word of $n-2$ occurrences written in a totally ordered alphabet A whose first letter is a. We trace the tree σ by successive grafts of the branches defined by the following instructions:

1. Let β_1 be the left factor of $a\ \sigma$, that is maximal without repetition of a letter. Let z_1 be the first absent letter of $a\sigma$. $\beta_1 z_1$ is the first branch.

2. Let σ_i be the right factor of $a\sigma$ not yet used. Let β_i be the left factor of σ_i maximal without having any letter, other than the first, appear in a branch already traced. Let z_i be the first letter of A which is absent from σ and the branches already drawn: $\beta_i z_i$ is the i-th branch.

3. If σ has another right factor not yet used, return to (2), otherwise graft the branches in the order of their first letter.

It is easy to show that:

- every point of the graft of a branch belongs to a branch of inferior rank in the succession of branches; and
- every letter of the alphabet other than the first appears once and once only in a branch without being a graft point.

We thus have come to the current graphic conception of a tree. It seems quite natural to give trees a new definition in terms of set mappings: every letter, except *a*, can be seen as being grafted to another once and only once (that is *the* branch where this letter is not the first).

1.1.3 Second definition of trees

Let A *be a finite set and* a *one of its elements. A tree rooted in* a *and joining all the elements of* A *is any orderable mapping* φ *of* $A - \{a\}$ *into* A. *By an orderable map we mean here a map to which we can associate a total ordering of* A, *beginning with* a, *and such that for every* x, *other than* a, $\varphi(x)$ *comes before* x *in this total order.*

Example:

$x \in A$	$a\ b\ c\ f\ d\ e\ g$
$\varphi(x)$	$a\ b\ a\ f\ a\ f$

(on the first line the elements x are written in the order of their total order)

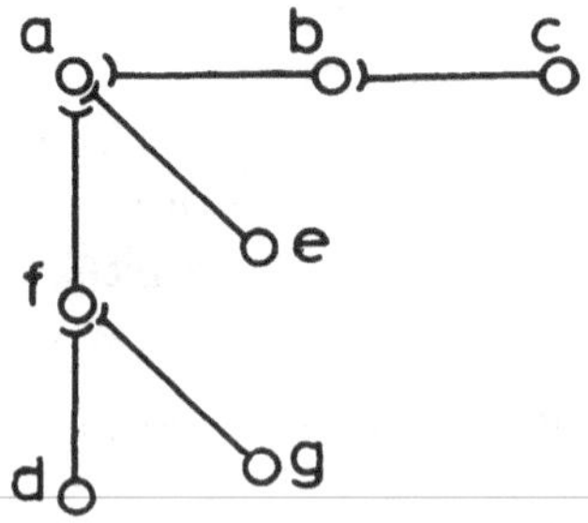

Fig. 7. *Tree associated with* abafaf

Note: The order associated with φ is that of the appearence of the letters in the branches of the word *abafaf* studied above.

The algorithm 'to trace a tree' defined above obviously generates such a function φ because any letter x other than a must:

– either appear in the word σ and then appear by its first occurrence in a word β_i in a rank superior to one in σ, or

– not appear in σ and in that case there must exist a branch i for which $x = z_i$ (each beginning of a branch is a repetition in the word σ, from whence the necessity of the letter being absent from σ and vice versa).

In every case, to the letter x can be associated as the image $\varphi(x)$ the letter which precedes x in the branch where x is not the first letter. $\varphi(x)$ is a letter that comes before x in the order of the first occurrence of the letters in the branches that follow.

1.1.4 Algorithm for coding a tree

Let us now procede to the coding of a tree. Consider a tree defined by definition 2. That is, let our object be a tree defined as an orderable mapping of the type defined above: $A - \{a\}$ into A.

1. In the order associated with φ take the first letter z_i, which is never an image, and take its image, the image of its image, etc. until a is reached. The sequence thus obtained is then written backwards, starting with a. This branch is called $\beta_1 z_1$.

2. In the order associated with φ, take the first letter z_i not yet chosen and which is not an image. Take its image, the image of its image, etc. until a letter is encountered that has already been written in a branch previously traced. Write the sequence so obtained in reverse, and call this branch $\beta_i z_i$.

3. If there still exists another letter z_i not yet taken, repeat step (2); otherwise write $\sigma = \beta_1 \beta_2 \ldots \beta_r$ where r is the number of branches obtained.

The 'round trip' is now completed. Starting with a 'word tree' of $n-2$ letters we have given an algorithm to construct an 'orderable mapped tree' (which can be easily identified with the drawn representation of a tree), and conversely, beginning with the second construction, we have given another algorithm to obtain the first.

Let us be explicit. The words of $n-2$ letters and the orderable mappings are in one-one correspondance. We are now in a position to consider the following theorem on trees.

1.1.5 Cayley's theorem

There exists n^{n-2} trees joining n given vertices. In fact, in an alphabet of n

letters the number of words having $n-2$ letters is n^{n-2}. For each of these words there corresponds by the 'tracing algorithm', an orderable mapping φ and for two distinct words there exists two distinct orderable mappings.

Conversely, for every orderable mapping φ there corresponds by the 'coding algorithm' a word of $n-2$ letters which generates φ by the 'tracing algorithm'. (In order to trace the corresponding tree, we take the alphabetic order associated with the orderable map used for coding).

Another exercice:
Given A: *a j m k l b c d e f g h i*

x	*a*	*j*	*m*	*k*	*l*	*b*	*c*	*d*	*e*	*f*	*g*	*h*	*i*
$\varphi(x)$		*a*	*j*	*j*	*k*	*m*	*m*	*k*	*l*	*l*	*l*	*k*	*j*

branches:
a j m b
m c
j k d
k l e
l f
l g
k h
j i

$\sigma = j\ m\ m\ j\ k\ k\ l\ l\ l\ k\ j$

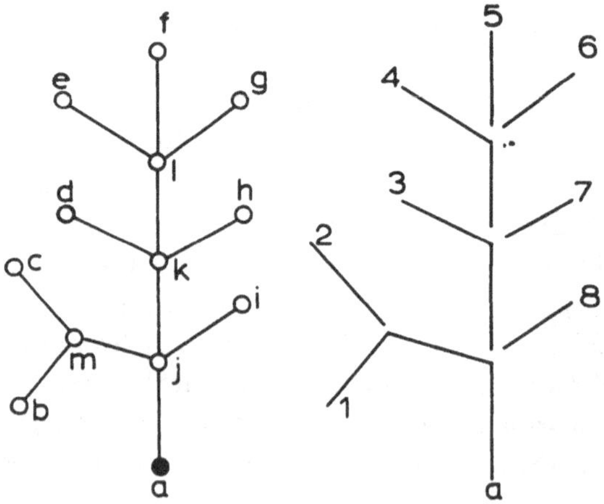

Fig. 8

1.2 Trees in Dyck words

We shall now consider trees from a different point of view. We are given lines and ask how they can be assembled into trees. The given lines are designated by letters and to distinguish the two extremities of a line, each line is given a direction (orientation). For n lines we have $2n$ extremities, some of which will be indistinguishable from one another, thus forming classes: exactly $n+1$ classes: the $n+1$ points of a tree having n lines.

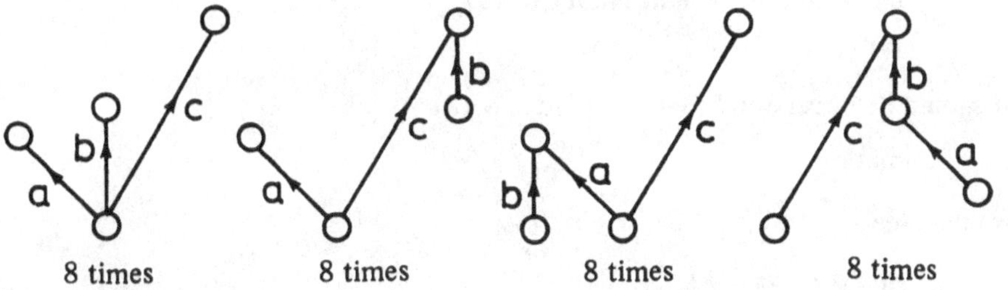

Fig. 9. *The herborium of 32 possible trees formed with 3 directed lines* A, B, C. ('8 times' indicates that we can associate 2^3 trees to each of the trees drawn above by only modifying the directions of the lines)

The essential idea to be developed here appears in Fig. 9 below where to code a tree we make a complete circuit, beginning from one of the points.

We first start on line C, this is written c;

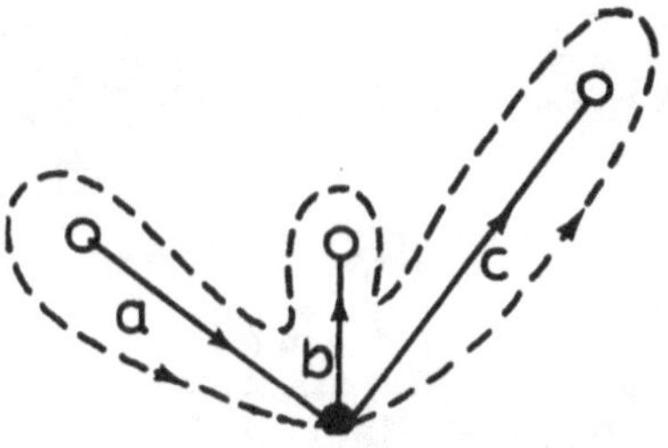

Fig. 10. *Tree cc'bb'a'a*

we begin again in the opposite direction on line C, written cc';

we then follow line B, in both directions which gives us now $cc'bb'$; and

finally, we follow line A in both directions thus obtaining $cc'bb'a'a$. Beware of A: the direction (a) is taken after its opposite direction (a').

When there is nothing left to take, then every line has been traced once in each direction and we have come back to the starting point. What kind of word has been written in this manner?

1.2.1 Dyck words

Let $\mathcal{A} = \{a,a',b,b', \ldots\}$ be a finite alphabet such that the letters are paired: a' is called the inverse of a and conversely. The inverse of any letter e is written e'.

We define the empty word $\wedge$ as the word having zero occurrences of letters.

For the words written in the alphabet $\mathcal{A}$ we will write the following equivalence relation:

$$aa' \sim a'a \sim \wedge \text{ (the empty word)}$$
$$bb' \sim b'b \sim \wedge, \ldots, \text{etc.}$$

It should be understood that if a word is written:

$$\sigma aa' \rho$$

we then have

$$\sigma aa' \rho \sim \sigma \rho$$

To recapitulate, this equivalence authorizes us to delete, to insert, or to concatenate a pair of adjacent and inverse letters anywhere in a word. A Dyck word is a word written in the alphabet $\mathcal{A}$ and is equivalent to the empty word (that is, it reduces to the empty word by successive deletions of pairs of adjacent and inverse letters). If a Dyck word does not contain a repetition of occurrences of the same letter it is called simple.

Example of simple Dyck words:

$a\ b\ c\ c'\ b'\ a'$ $\qquad$ $a\ b'b\ a'c'c$

$a\ b'\ b\ c\ c'\ a'$ $\qquad$ $a\ a'\ b\ b'\ c\ c'$

Algorithm for simple Dyck words of a tree

We are given a tree whose lines are arbitrarily directed. To each line A we assign two inverse letters, one a which corresponds to the direction of A in the direction of its orientation, the other a' in the reverse sense. We arbitrarily choose a first letter.

After the first chosen letter, write the adjacent letters of the tree for as long as possible following the rules:

1. do not write the same letter twice; and
2. at any point x in the algorithm, if possible, choose a letter whose inverse has not yet been written.

Example:

15′22′3′344′56788′7′6′1′

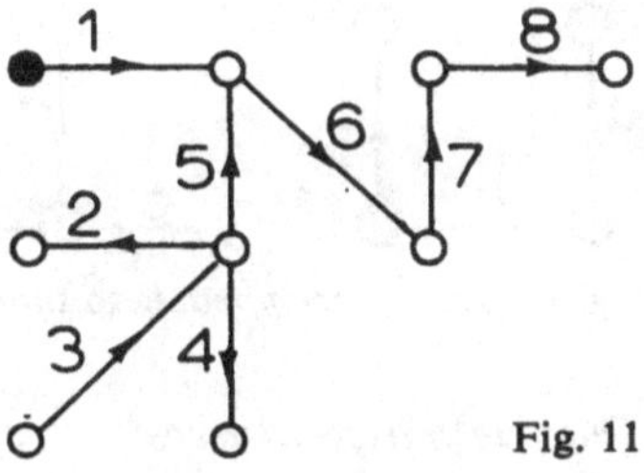

Fig. 11

Example of an application: We are asked to find the shortest word in the tree that begins with 3 and ends with 8. From the word above we extract the factor:

344′5678

and simplify to

35678

Defining trees in terms of simple Dyck words turns out to be useful in certain applications.

It is simple to show that the algorithm above generates a simple Dyck word such that:

- it contains all the letters of the tree:
- two adjacent letters are adjacent in the tree; and
- the last letter is adjacent to the first.

To be more exact $\prod_{x \in X}((v(x)-1)!)$ words are possible given the freedom of choice that exists for every point x. The valence $v(x)$ (or degree) of the point x is the number of lines containing x as an extremity.

Note for the proof: For each point x, except the starting point, there exists as many letters in the direction 'away' from x as in the direction 'towards' x. The inverse of the first letter to appear in the first direction above will be the last letter in the second direction.

1.2.2 Definition of a tree by Dyck words

Consider the inverse of what we have just done with a simple Dyck word: it will correspond to a unique tree. Let us reconsider the four previous examples which we will now write as follows:

$a\,b\,c\,c'\,b'\,a'$ $\quad a\,b'\,b\,c\,c'\,a'$ $\quad a\,b'\,b\,a'\,c'\,c$ $\quad a\,a'\,b\,b'\,c\,c'$

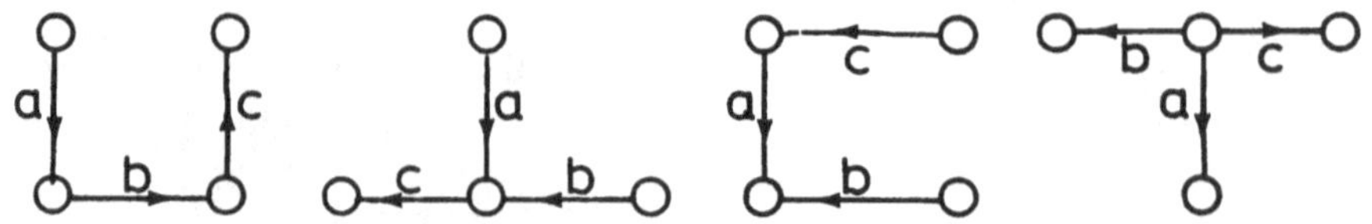

Fig. 12. *Trees corresponding to simple Dyck words*

Third definition of trees
Given an alphabet of 2n *letters where all the letters are paired and where two letters of the same pair are called inverses. We define a tree as any simple Dyck word written in this alphabet. That is, all words containing each letter at most once, that reduce to the empty word by successive deletions of letter pairs that are adjacent and inverse.*

This definition is equivalent to definition 2 given previously for trees, as we will now show.

Consider a simple Dyck word σ written in the alphabet previously given. Let two inverse letters of the alphabet correspond to a line (beginning in either direction) and suppose that each letter e, touches two points: $l(e)$ its left point and $r(e)$ its right.

For two adjacent letters in the word σ such as ef we can write:

$$d(e) = g(f)$$

and for two inverse letters e and e':

$$d(e) = g(e') \quad \text{and} \quad g(e) = d(e')$$

It follows that the $2n$ letters of the word σ define $n+1$ distinct points which are the left point of the first letter of the word (which is the root), and the right point of each letter written whose inverse has not yet been written.

n lines; n+1 points and a root; we need only further observe that the mapping which associates for each letter e whose inverse has not yet been written, the right point $r(e)$ with the left point $l(e)$, a point it should be noted, whose image has already been found, is none other than the orderable mapping φ in definition 2.

We might ask what are all the simple Dyck words which correspond to the same tree. We need only agree to the following equivalence in σ:

$$\lambda\mu \sim \mu\lambda$$

where γ and μ are Dyck words and $\gamma\mu$ is a factor of σ.

1.2.3 Three new definitions of a tree

Given a tree, let U be its set of lines and X its set of points. The lines may be arbitrarily directed. We shall consider a mapping

$$I: U \rightarrow X,$$

which associates each line to its initial extremity (relative to its orientation) and the mapping

$$T: U \rightarrow X,$$

which associates each line to its terminal extremity (relative to its orientation).

Consider a point r, a simple Dyck word of the tree rooted at r, and once again the sequence of letters for which the inverse has not yet been written. The mapping ψ which associates the line of a letter e of this sequence to the right of this letter is;

1. a bijection of U on $X - \{r\}$;
2. compatible with I and T in the sense that the image of a line is the extremity of that line; and
3. an orderable mapping in the sense (very special) that the order extremity of the line has already been found. That is, it is either r, or the image of a line previously considered in the sequence.

These three properties are obviously properties of Dyck words. They lead to a new definition of trees.

Fourth definition of trees
Let X *be a finite set of points and* U *a finite set of lines. Let* $\mathrm{I} : \mathrm{U} \rightarrow \mathrm{X}$ *and* $\mathrm{T} : \mathrm{U} \rightarrow \mathrm{X}$ *be two mappings such that the line* u(u$\in$ U) *joins the point* I(u) *to the point* T(u), *its two extremities.*[2] *We let* r *be any arbitrarily chosen point.*

The quadruplet (X, U, I, T) *is a tree if there exists a mapping* ψ: $\mathrm{U} \rightarrow \mathrm{X} - \{\mathrm{r}\}$ *which is:*

1. *bijective;*
2. *such that* ψ(u) *is an extremity of* u; *and*

2. We have thus a quadruplet, that is, a list of our mathematical objects in a predetermined order. The quadruplet (*X, U, I, T*) which has just been defined is a *graph*. A precise definition of graphs can be found in section 3.1.1.

3. *such that there exists an order in* U, *for which the extremity of* u, *other than* $\psi(u)$, *is* r, *or the image of a line inferior to* u.

It is easy to show that this definition is equivalent to the three preceding ones. We can now deduce without difficulty, a fifth and sixth equivalent definition in terms of words.

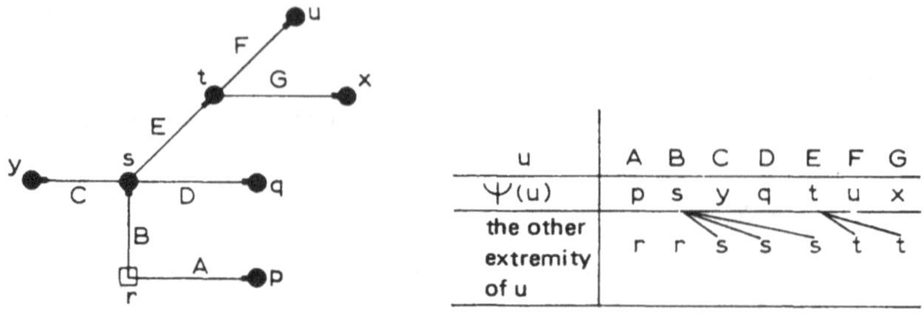

Fig. 13–14. *Mapping ψ characterising a tree*

Fifth definition of trees

By a tree, we mean a quadruplet (X, U, I, T) *where* X *is a finite set (of points),* U *a finite set (of lines), and* I *and* T *are mappings from* U *into* X *such that from every point to every other there exists a unique minimal word.*

To define the words of a quadruplet (X, U, I, T) we associate with each line with extremities x and y two inverse letters a and a', such that the left point of a and the right point of a' is x, and the right point of a and the left point of a' is y. We write the letter e immediatly followed by f if, and only if, the right point of e is equal to the left point of f as above for the Dyck words.

Sixth definition of trees

A quadruplet (x, I, T), *where* $U = |X| - 1$,[3] $I : U \to X$ *and* $T : U \to X$ *and such that from one point to another there always exists a word, is called a tree. We shall also define a tree as a connected*[4] *quadruplet with one more line than points.*

3. The cardinal of a set A written $|A|$ or sometimes card (A) is the number of elements in this set.

4. A precise definition of connectivity will be found in 3.1.1 equivalent to the following: from one point to another there always exists a word.

1.2.4 The Hamiltonian tree theorem

Hamilton's famous problem is that of finding a circular permutation of given elements compatible with a certain relationship between objects defined in advance.

Consider a tree (we already know six equivalent definitions). It has three sets of elements: letters, lines and points.

Now, for example, let us attempt to number the points of the tree in Fig. 14 by making jumps that are as short as possible. A jump of two points appears to be inevitable. Do trees exist for which jumps of three points are necessary?

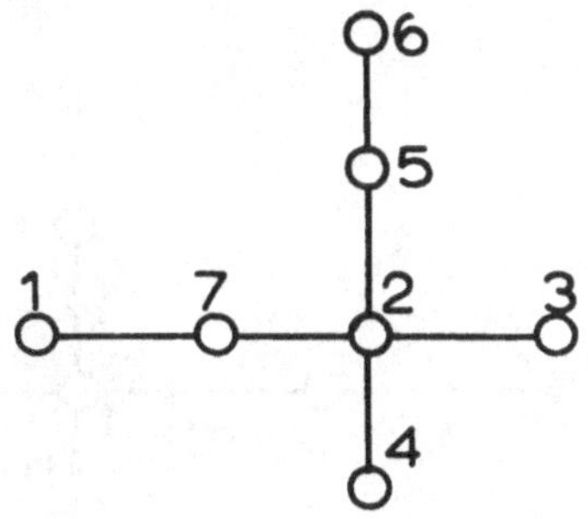

Fig. 15. Circular permutation of the points of a tree with jumps of no more than two points

Hamilton's theorem

T1. *For letters* it is always possible to circularly number the letters of a tree in such a way that each letter and its successor in this enumeration will be adjacent (that is, that the right point of the first will be the left point of the second).

T2. *For lines:* lines of a tree can always be circularly numbered in such a way that between each line and its successor there exists 0 or 1 line.

T3. *For points:* the points of a tree can always be circularly numbered so that between any point and its successor there exists 0, 1 or 2 points.

T1 is proved by the algorithm for the simple Dyck word of a tree given in section 1.2.1. T2 is proved by treating a Dyck word with the following algorithm.

Algorithm for the Hamiltonian numbering of lines

1. In a simple Dyck word of a tree, recopy the letters of odd rank.

2. Replace each remaining letter by the line to which it corresponds.

We should first note that in a simple Dyck word, the inverse letters have parities of distinct rank. In the numbering obtained, all the lines are thus numbered, and two successively numbered lines are separated by at most one other, namely the line corresponding to the deleted letter, if it is not identified with either one of these two.

Examples: We are given the tree below (Fig. 16). We construct a simple Dyck word, σ, and apply the Hamiltonian algorithm numbering of the lines.

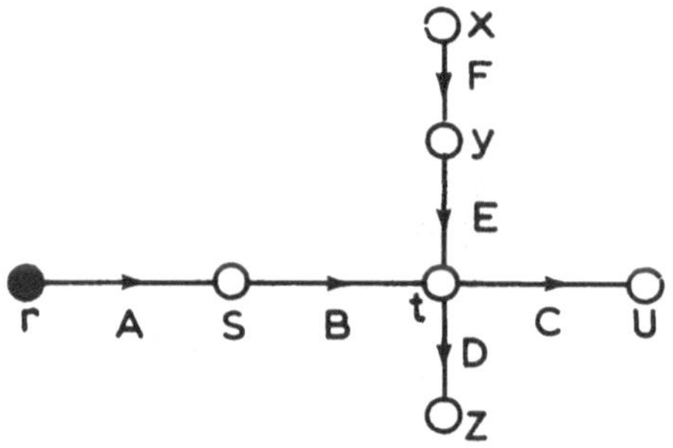

Fig. 16

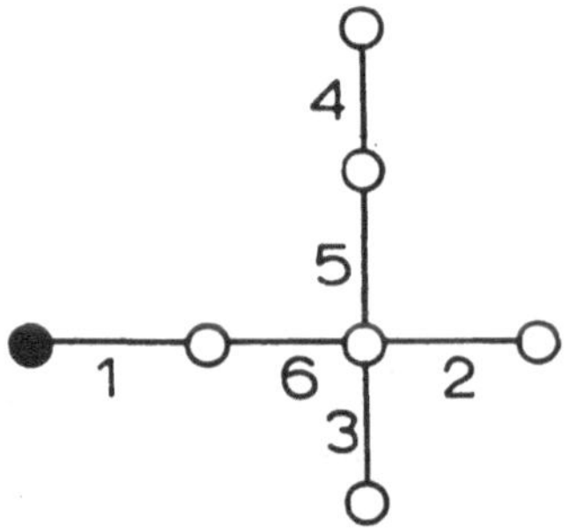

Fig. 17

$$\sigma = a\ b\ c\ c'\ d\ d'\ e'\ f'\ f\ e\ b'\ a'$$

instruction 1: $a\ \ c\ \ d\ \ e'\ \ f\ \ b'$
instruction 2: $A\ C\ D\ E\ F\ \ B$

T3 is proved by treating the Dyck word of the tree with the following algorithm.

Algorithm for the Hamiltonian numbering of points

1. In a simple Dyck word σ of a tree, recopy the letters of odd rank.
2. To each recopied letter e associate its left point if e precedes e in σ, its right point if e precedes e' in σ.
3. At the head of the list thus obtained, place the left point of the first letter of σ.

Example:

$$\sigma = a\, b\, c\, c'd\, d'\, e'\, f'\, f\, e\, b'\, a'$$

instruction 1:	a	c	d	e'	f	b'
instruction 2:	s	u	z	y	x	t
instruction 3:	$r\, s$	u	z	y	x	t

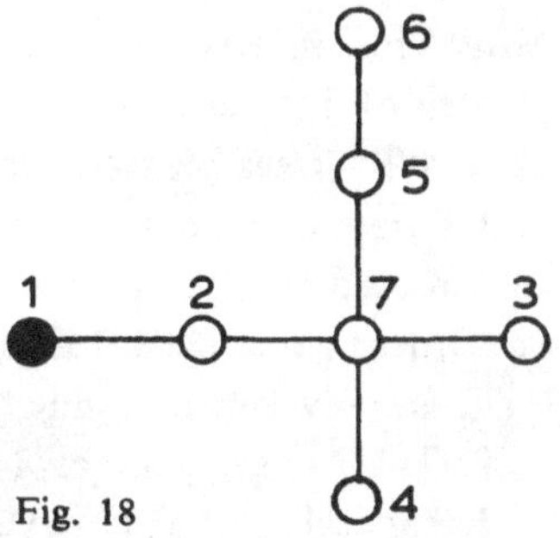

Fig. 18

The above algorithm leads to a numbering of all the points because it defines the *ordered* mapping ψ given in definition 4 for trees. As for the jumps from one point to its successor in the numbering, they were at most one for the line ordering and thus at most two for the extremities of these lines.

2. Rooted semi-lattices

2.1 Rooted trees and partial order

Définition: A rooted tree with the root r is a connected graph, $A = (X, U, I, T)$ where T is an injection whose image set is $X - \{r\}$. The elements of U, called arcs by some authors, can be represented in this case by the ordered pairs (Iu, Tu) without ambiguity.

Exercise: Every vertex is a *descendant* of the root. Said another way, for $x \in X$, either $(r, x) \in X$, or there exists a path $(r, z_1)\,(z_1, z_2), \ldots, (z_p, x)$ each element of which is an element of U.

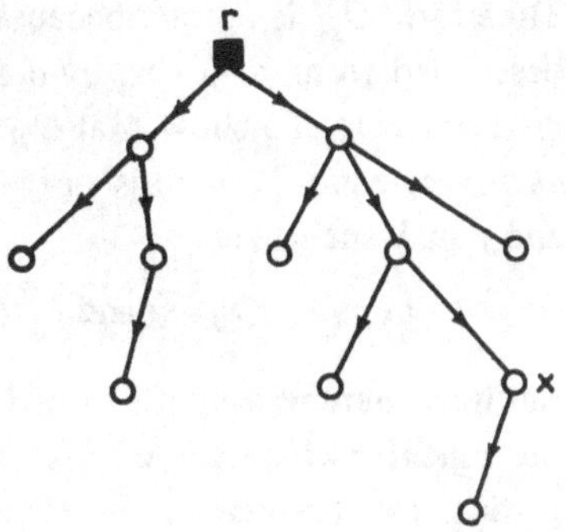

Fig. 19

Note: Trees in 'free words' defined in 1.1 are, in fact, rooted trees, since they are planted on the vertex that initially served for their coding. Moreover, since there exists n^{n-2} trees planted on the root r, and n possible roots, there exists n^{n-1} rooted trees on n points.

A rooted tree can be made to correspond to an order O_A on X (see chapter 2 on 'Ordering and classification') that is, a certain set of ordered pairs O_A: $(x, y) \in O_A$ is equivalent to saying that $x \leqslant y$. O_A is defined as follows:

1. $U \subset O_A$: every arc of A is an element of O_A.
2. For each x of X, $(x, x) \in O_A$; (O_A is reflexive).
3. For every $(x_1, x_2) \in O_A$ and $(x_2, x_3) \in O_A$, $(x_1, x_3) \in O_A$ (O_A is transitive).
4. The only elements of O_A are those deduced from (1), (2) and (3).

We say that O_A is obtained by the reflexive and transitive closure of U. Thus, the rooted tree of Fig. 20 generates the *relation* for which the *network* is that of Fig. 21.

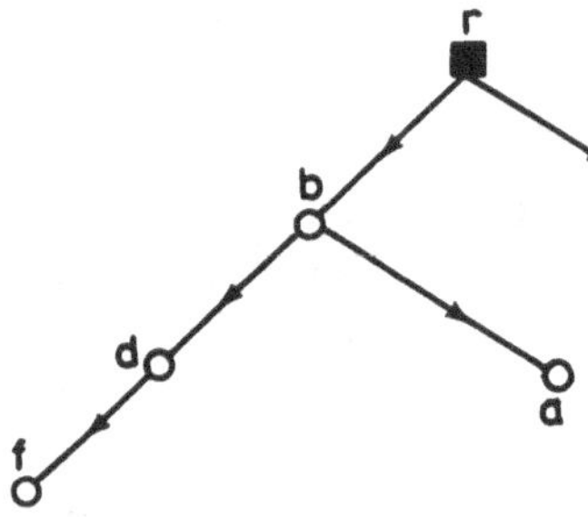

Fig. 20

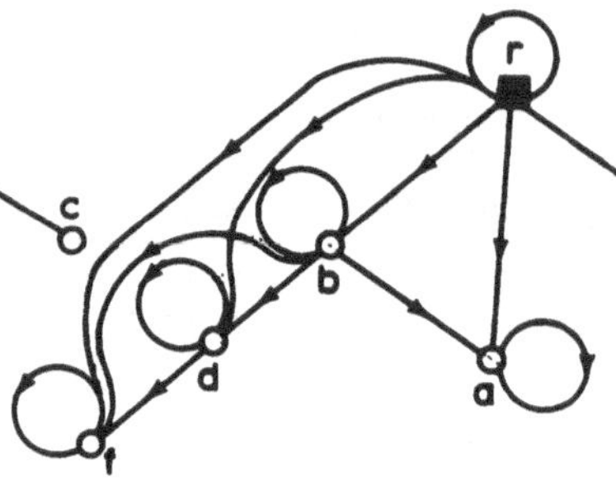

Fig. 21

Therefore, O_A is unique because (x, y) is an element of O_A if, and only if, y is descended from x. O_A is, by definition, a reflexive and transitive relation: it is a *quasiordering.* To show that O_A is really an order, we must show that O_A is also *anti-symmetric.* Were this not the case there would be two distinct elements x and y in X such that

$$(x, y) \in O_A \quad \text{and} \quad (y, x) \in O_A$$

But in A there was a path from x to y and another from y to x which contradicts the injective character of I as well as the existence of a root for which all the vertices are decendants. Therefore, O_A is an order. The order O_A associated with a rooted tree has the following properties:

1. Each distinct element of the root 'covers' another unique element in the technical meaning of order because here:

$$y \text{ covers } x \Leftrightarrow (x, y) \in U,$$

or T is an injection.

2. Given two elements x and y, the set of elements greater than x and y has a minimum.

If one of these two elements is r, r is that minimum, otherwise x covers x_1 which covers x_2, if $x_1 = r$, etc. Therefore there exist two sequences of covers: one from x to r, the other from y to r. As a consequence of (1), if the same letter appears in the two sequences, then the two sequences are identical to the right of this letter. The maximum subset (reduces at most to r) thus constitutes a set of elements greater than both x and y and the first letter common to the two sequences (r for example) is its minimum.

Example:

$$f\,d\,b\,r$$
$$a\,b\,r$$

thus

$$f \wedge a = b$$

similarly

$$f \wedge c = r$$
$$a \wedge b = b$$
$$a \wedge r = r$$

O_A is a *semi-lattice.* This type of semi-lattice is discussed in chapter II 'Ordering and Classification' under the name *trees.*

2.2 Reorientation of the lines of a tree

Let A be a tree (X, U, I, T). If r is a point of A, we can orientate (direct) or reorientate the arcs of U to obtain a rooted tree A_r of root r. We will let U_r stand for the set of arcs of A_r.

This reorientation is obtained by applying two simple rules:

1. If $I(u) = r$ & $T(u) = x$, or if $T(u) = r$ & $I(u) = x$, then $(r, x) \in U_r$.

2. If $(x, y) \in U_r$, and if $I(u) = y$ & $T(u) = z$, or $T(u) = y$ & $I(u) = z$, then $(y, z) \in U_r$.

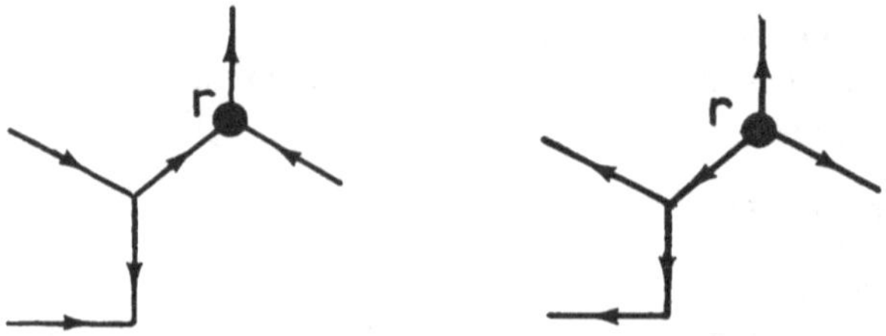

Fig. 22 *Tree A* *Rooted tree A_r*

We need only refer to the algorithm for the simple Dyck words of A (in 1.2.1) to see at once that A_r is a rooted tree of root r and that the reorientation of A in a rooted tree of root r is unique.

This suggests a new definition of trees.

Seventh definition of trees
A tree is any graph identified with a rooted tree after a change in direction of one part of its arcs.

The definition of rooted tree is given in section 2; it is easy to relate this seventh definition of trees with definition 4 above.

2.3 Various uses of rooted trees

A large number of configurations of situations can be represented by rooted trees. A few examples will be given below.

2.3.1 Rooted trees and classification

One of the usual ways of classifying the objects of a set is to define a series of partitions on the set which are made finer and finer (nesting partitions). Thus, in zoology, the set of all animals is divided into branches, classes, orders, families, down to the individuals themselves.

An example of this type of organization is this book which is first divided into articles that are themselves subsequently subdivided into sections and subsections. The rooted tree of Fig. 23 illustrates the organization of this article.

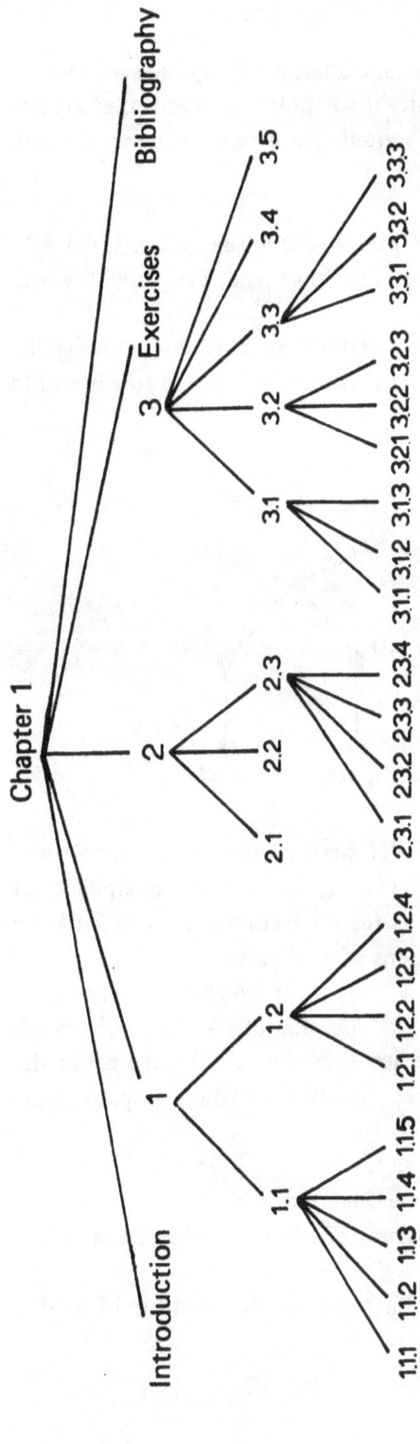

Chapter 1
Introduction
1
2
3
Exercises
Bibliography
1.1
1.2
2.1
2.2
2.3
3.1
3.2
3.3
3.4
3.5
1.1.1 1.1.2 1.1.3 1.1.4 1.1.5
1.2.1 1.2.2 1.2.3 1.2.4
2.3.1 2.3.2 2.3.3 2.3.4
3.1.1 3.1.2 3.1.3
3.2.1 3.2.2 3.2.3
3.3.1 3.3.2 3.3.3

Fig. 23

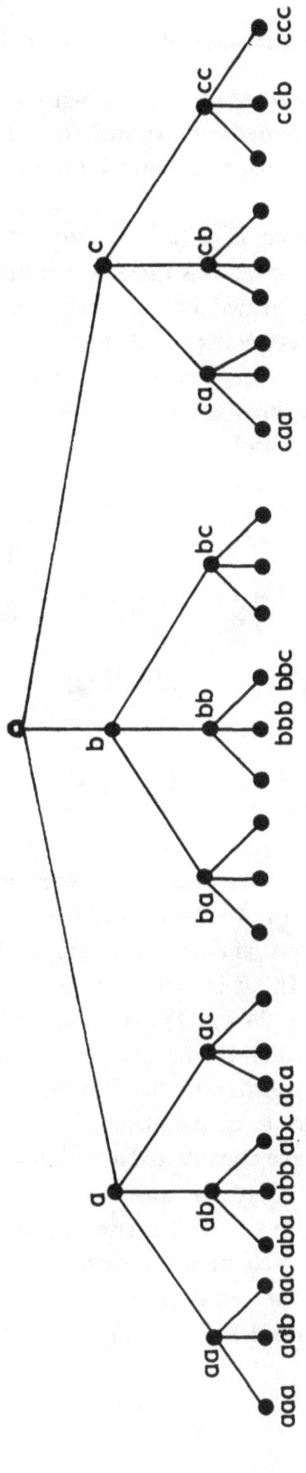

a
b
c
aa
ab
ac
ba
bb
bc
ca
cb
cc
aaa aab aac aba abb abc aca
bbb bbc
caa
ccb ccc

Fig. 24

2.3.2 Exponential trees, factorial trees

The set of all words written in a finite alphabet F of cardinal n, can be represented by a rooted tree. It will be finite if we limit it to words that are less than equal to a finite k for example. This rooted tree is said to be exponential.

Example: $E = \{a, b, c\}$ $n = 3$, $k = 3$
Each vertex has three successors: there are three words of one letter, $3 \times 3 = 3^2$ words of two letters, 3^3 words of three letters, etc. In general, there are n^k words of k letters in an alphabet of n letters.

Another classic type of rooted tree is obtained by only constructing the words in which the same letter does not appear twice. This rooted tree is said to be *factorial.*

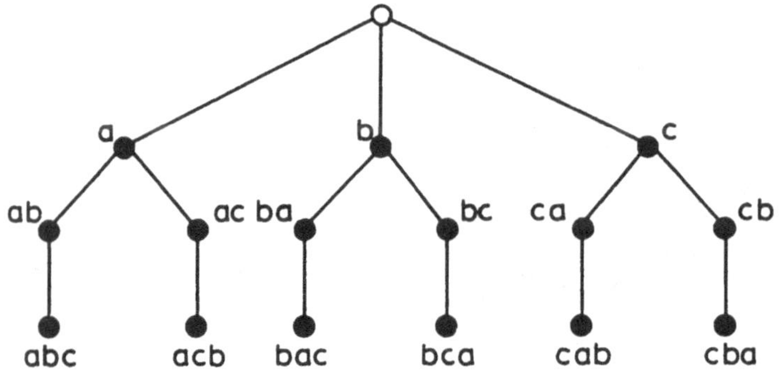

Fig. 25

Each vertex has one successor less than its predecessor. There are $n(n-1)\ldots(n-k+1)$ words of length k without repeating the letters in an alphabet of n letters, and $n! = n(n-1)\ldots 2.1$ such words for $k = n$ (thus six words for $k = n = 3$). This is in fact the number of permutations of n objects.

Figures 24 and 25 have been drawn with one convention: an arc representing the adjunction of the letter a is further to the left than that of the letter b, which is itself further to the left than that of the letter c. We can therefore use the properties of the plane of the sheet of paper to define a particular word of the root tree constructed as follows:

1. start with the root;
2. never follow the same arc twice in the same direction; and
3. at each vertex choose the arc that has not yet been traced and which is farthest to the left.

This word is a particular case of the words obtained by the algorithm of section

1.2.1. The difference between the two is due to the restrictions that make the word defined above unique.

Let us consider the order in which the vertices are encountered for the first time. For the tree in Fig. 24, this order is *a, aa, aaa, aab, aac, ab, aba, . . ., ccb, ccc*. For the tree in Fig. 25 it is *a, ab, abc, . . ., cb, cba*. In either case it is the *Lexicographic* order associated with the *alphabetic order a, b, c*. The lexicographic order is used in the quasitotality of dictionaries.

2.3.3 Generative tree and parenthesization

Many linguists perceive the structure of a sentence through the use of trees. The sentence is subdivided into nesting blocks. We shall borrow an example from Gross and Lentin[5] : *The child picks an apricot.*

SE: sentence
NG: nominal group
VG: verbal group
AR: article
NO: noun
VE: verb

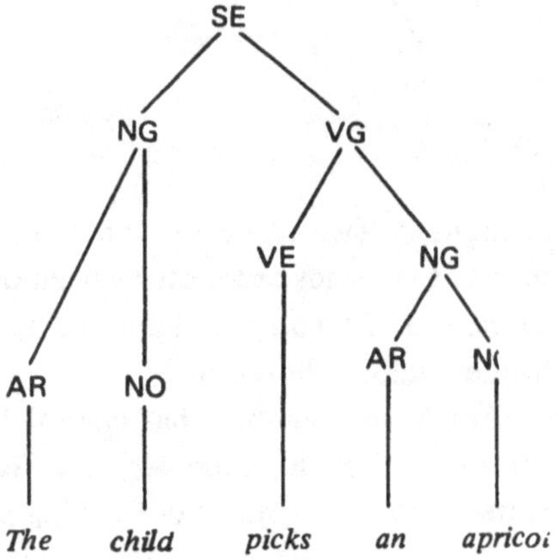

Fig. 26

The rooted tree obtained, a *generative tree*, cannot be anything but unique. It corresponds to a system of parenthesization of the sentence:

(the child) (picks (an apricot)).

Another system of parentheses would generate another rooted tree.

2.3.4 Optimization by recurrence on decision trees

In many situations we are led to consider a set of linked states, or situations related to one another, in such a fashion that the set is easily described by a

5. See bibliography.

rooted tree. This is a particular case of a large number of games and decision problems.

Example 1. A traveller can reach a nonmaritime city by plane (A). He could also take a boat (B) and finish his trip by car (C) or railway (D). The set of these possibilities can be summarized by a rooted tree.

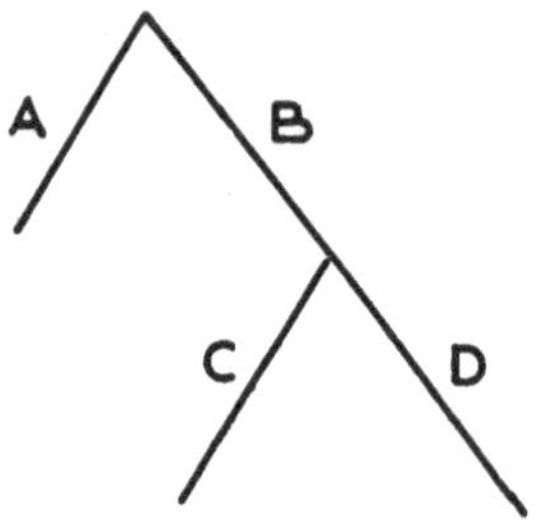

Fig. 27

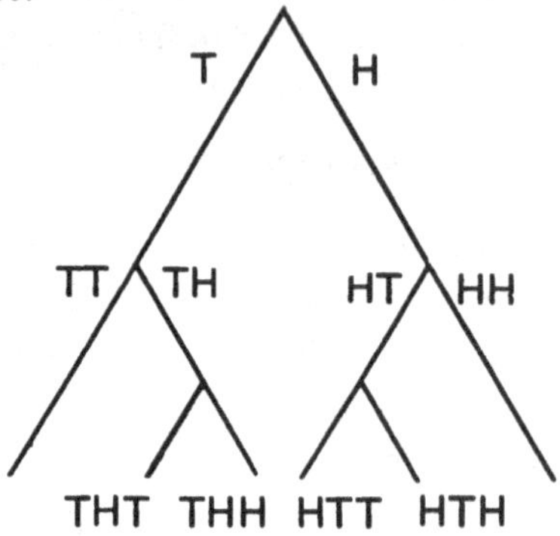

Fig. 28

Example 2. Two players toss a coin, H for heads, T for tails. The game ends as soon as two heads or tails have been obtained, not necessarily consecutively.

Example 3. Two players A and B play as follows. A chooses 1 or 2. B adds 1 or 2 to the number chosen by A and so on. The game ends when one of the players reaches 5, the number that cannot be exceeded. Each player in turn is placed before a problem of choice. The set of outcomes of possible games can be represented by the set of descending pathways in the following rooted tree.

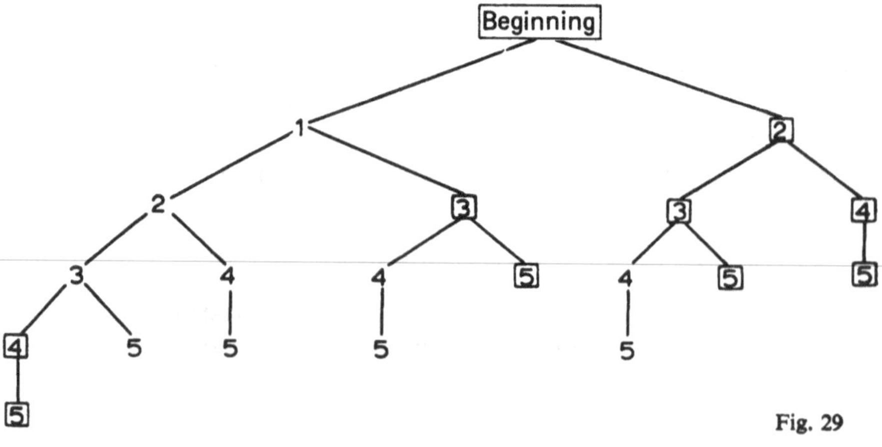

Fig. 29

In this sort of problem, the study of optimum choice is not made starting from the root, but rather from the vertices that have no successor and working back towards the root. Also, in example 1, of the traveller wishes to make his trip as economically as possible, he should first compare the cost of the car and the railway, and then add the *lowest* of these two alternatives to the cost of the boat in order to make his choice between the various conveyences and the plane.
In example 3, suppose that the winning player is the one who gets 5 first. Put a box around the vertices having no successors and which represent a victory for A. The other five vertices represent a victory for B. By working back up the appropriate side of the tree (taking into account the name of the player whose turn it is) it is possible to partition all of the vertices of the rooted tree. That is, it is possible to partition all of the situations found in the game into a set of situations favourable to A (boxed) and a set of those favourable to B. The beginning having been boxed, A has the ability to win without fail if he does not make an error (by playing according to the stressed arcs).

Such recurrent optimization reasoning had already appeared in Pascal's treatise on *'Triangle arithmetique'*. It has been more recently given a name related to modern economics: *dynamic programming*.

3. Modules on trees

3.1 Trees of a graph

3.1.1 Definition: A quadruplet (X, U, I, T) is called a graph G where

X is a set of the vertices of G;
U is a set of the arcs of G;
$I : U \to X$ and $I(u)$ is called the initial extremity of u; and
$T : U \to X$ and $T(u)$ is called the terminal extremity of u.
If X and U are finite, we can write $|X| = n$ and $|U| = m$.

We shall give an example of a graph that will be useful later.

$G = (X, U, I, T)$ where
$X = \{x, y, z, t\}$ $\quad n = 4$
$U = \{a, b, c, d, e, f, g, h\}$ $\quad m = 8$

I and T are given by the following diagram:

u	a	b	c	d	e	f	g	h
$I(u)$	x	y	y	x	z	t	x	t
$T(u)$	z	t	x	z	y	y	x	z

This graph is represented by the following figure:

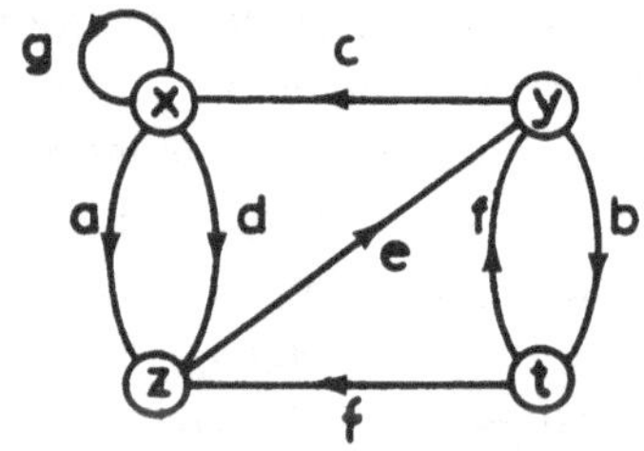

Fig. 30

The partial graph $G(A)$ of G where $A \subset U$ is the graph G after all the arcs not in A have been erased. We shall call the *boundary of an arc U* the sum of the extreme vertices of u and write it ∂u:

$$\partial u = I(u) + T(u)$$

The addition to be discussed here acts on elements of the same set; also it is commutative, has a zero term and each element is its own inverse ($x + x = 0$). Such are our conventions.

What then do we call the boundary of a sum of arcs? By definition the sum of their boundaries:

$$\partial(u_1 + u_2 + \ldots u_r) = \partial u_1 + \partial u_2 + \;\ldots\; + \partial u_r$$

Example:

$$\partial a = x + z$$

$$\partial c = x + y$$

$$\overline{\partial(a+c) = x + x + y + z = 0 + y + z = y + z}$$

The notion of boundary brings us to a new definition of trees.

Eighth definition of trees
A tree is a graph for which the equation of the boundary $\partial z = A$ *has a unique solution in* Z *for all subsets of even cardinality of the vertices.*

This definition can be reduced without difficulty to definition 5 for trees. For this we need only observe that the boundary of the edges of a word is none other than the sum of the two extremities of this word.

A graph G is said to be *connected* if each pair of vertices is the boundary of a sum of arcs. In the example we have:

$$\partial(a+d) = 0$$
$$\partial(c+d+e) = 0$$
$$\partial(a+d+c+d+e) = 0$$

Thus there exists sums without a boundary.

We call a *cycle* of G all arc sums having a zero boundary. The sum of two cycles is another cycle.

3.1.2 Ninth definition of trees

A tree V *of a graph* G *is any subset* V *of its set* U *of arcs (or sums of arcs) not including cycles other than O, and maximal with respect to this property.* (Note: If V is a tree of G, $G(V)$ is not necessarily connected.) *A 'tree graph' is any connected graph (X, U, I, T) for which* U *is a tree.*

The equivalence of definitions 8 and 9 can be shown in two steps:
– the existence of a solution to the equation $\delta Z = A$ (for A of even cardinality) is equivalent to connectivity; and
– the uniqueness of a solution is equivalent to the absence of all cycles other than zero.

For this second point, in fact:

$$[\partial z_1 = A \quad \& \quad \partial z_2 = A] \Leftrightarrow \partial(z_1+z_2) = A+A = 0$$

Let G be a connected graph and W a subset U such that $G(W)$ is acyclic. If card $W = n-1$, W is a tree. If card $W < n-1$, $G(W)$ is not connected but there exists in G arcs that join the connected components of $G(W)$, since G is connected. Let u be such an arc: $G(W \cup \{u\})$ is acyclic and card $(W \cup \{u\})$ = card (W)+ 1.

This complementation could be extended until a tree G has been obtained;

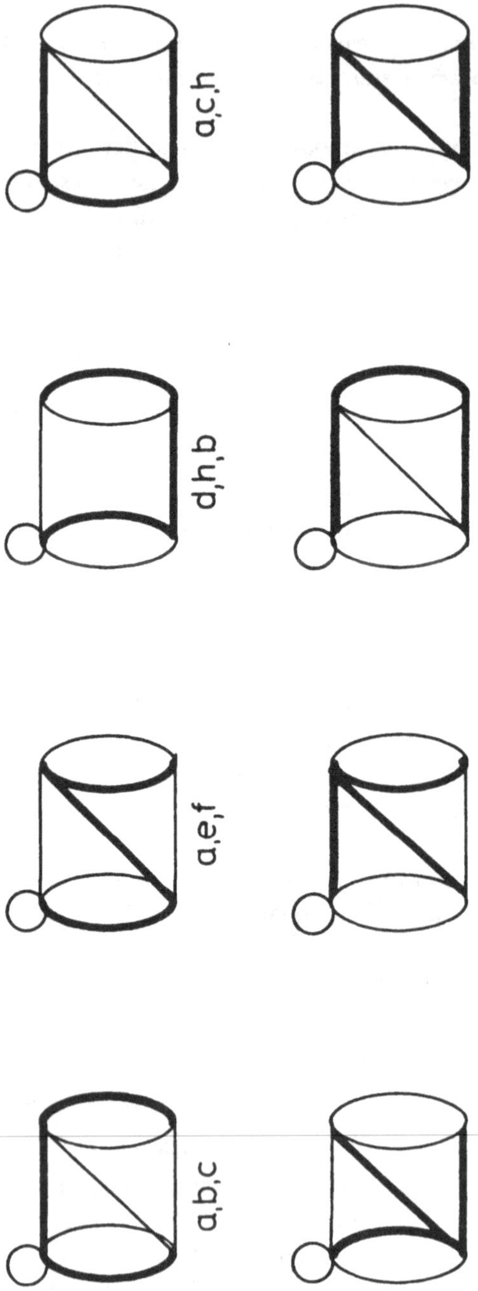

Fig. 30. *Eight of the twenty-one trees of the graph of Fig. 30 (The other thirteen can be found by exchanging arcs* a *and* d *or* b *and* f)

from this we get two propositions:
- every connected graph includes at least one tree as a partial graph; and
- in a connected graph, every set of arcs without cycles is contained in at least one tree.

3.1.3 Trees and linear independence

Suppose G has no loop (in the graph in Fig. 31 arc g is a loop). Let $\mathcal{L}$ be the family of subsets of U which do not contain a cycle. We have:

1. $\forall u \in U, \{u\} \in \mathcal{L}$
2. $V \in \mathcal{L}$ and $W \subset V \Leftrightarrow W \in \mathcal{L}$
3. $\forall U' \subset U$, the maximal elements of $\mathcal{L}$ in U' all have the same cardinality.

The maximal elements of $\mathcal{L}$ in U', that is, the subsets without cycles of U' that are not contained in any other subset are the trees of $G(U')$. Property (3) is thus proved. Property (2) expresses the fact that every subset of a set of acyclic arcs is itself acyclic.

The families of subsets which satisfy the axioms (1), (2) and (3) have a number of properties. We say that the ordered pair $(U, \mathcal{L})$ is a matroid for which $\mathcal{L}$ is the family of independent subsets. The independent subsets in a vector space also satisfy (1), (2) and (3). Bases in vector spaces (maximal independent subsets) correspond to trees (subsets without maximal cycles). It is for this reason that (1), (2) and (3) (or axioms equivalent to them) are sometimes said to be the axioms of linear independence.

The relation between trees and linear independence will become apparent in the following paragraph.

3.2 Flows and tensions

3.2.1 Flows in a graph

In many problems we are led to assign a number to each arc (respectively vertex) of a graph, which is then said to be valued. We are thus given a mapping from U (respectively from X) to a set of numbers.

Thus if $U = \{a, b, c, d, e\}$
and if $\nu: U \to \mathbf{Z}$ is defined by:

$$\nu(a) = 3; \nu(c) = -2; \nu(d) = 1; \nu(b) = \nu(e) = 0,$$

then the map can be represented in the usual way by a vector $\nu = (3, 0, -2, 1, 0)$.

As usual let **Z** be the set of relative integers:

$$\mathbf{Z} = \{\ldots, -2, -1, 0, 1, 2, \ldots\}$$

We can also write:

$$\nu = 3a - 2c + d$$

This latter form is more convenient and we shall use it here. ν is an element of $\mathbf{Z}^U$, the set of maps of U into **Z**.[6] We represent a as the vector (1, 0, 0, 0, 0) and similarly with b, c and d,

The elements of $\mathbf{Z}^U$, can be added. If $\mu = 2a - 3b + 2d$, then

$$\begin{aligned} \nu + \mu &= (3a - 2c + d) + (2a - 3b + 2d) \\ &= 5a - 3b - 2c + 3d \end{aligned}$$

or they can be multiplied by an integer

$$2\nu = 2(3a - 2c + d) = 6a - 4c + 2d$$

Division by an integer is not always possible, but we will not need this operation for our calculation. $\mathbf{Z}^U$ is thus a non-divisible abelian group (see Chapter VI, 'Measure Scales'). Expressions such as $3a - 2c + d$ can also be considered as elements of a commutative free group generated by U (see Chapter V, 'Monoids and Groups').

With multiplication by integers, $\mathbf{Z}^U$ is a module on **Z**. We have, in fact:

$$\begin{aligned} (Z_1 + Z_2)\nu &= Z_1\nu + Z_2\nu \\ Z_1(\nu + \nu') &= Z_1\nu + Z_1\nu' \\ Z_1(Z_2\nu) &= (Z_1Z_2)\nu = Z_1Z_2\nu \\ \forall Z_1, Z_2 \in \mathbf{Z} \quad & \& \quad \forall \nu, \nu' \in \mathbf{Z}^U \end{aligned}$$

In this paragraph, we will consider the elements of Z^U. We could, however take another set of numbers other than Z with a structure weaker or stronger: the field of rational numbers, for example. $\mathbf{Q}^X$ et $\mathbf{Q}^U$ are vector spaces on **Q**. The properties of modules are satisfied, and further, the division by a rational number is always possible. One can also take a non numerical structure, an abelian, for example. (We shall see such an example in section 3.3.2.)

6. If A and B are two sets, we write A^B as the set of mappings from B into A. Recall also that if A and B are both finite (here **Z** was not) we have the following equality:

$$|A^B| = |A|^{|B|}$$

We shall now define the boundary ∂u of an arc by

$$\partial u = T(u) - I(u),$$

and for any element of $\mathbf{Z}^U$, the boundary map is a morphism of $\mathbf{Z}^U$ into $\mathbf{Z}^X$:

$$\partial(Z_1\mu+Z_2\nu) = Z_1\,\partial\mu+Z_2\,\partial\nu$$

where Z_1, $Z_2 \in \mathbf{Z}$, $\nu\ \mu \in \mathbf{Z}^U$. The boundary can be calculated from its trees. Take the graph in Figure 30:

$$\begin{aligned}\partial(2a+e+b) &= 2\partial a+\partial e+\partial b\\ &= 2(z-x)+(y-z)+(t-y)\\ &= -2x+z+t\end{aligned}$$

An element φ of $\mathbf{Z}^U$ is called a *flow* when

$$\partial\varphi = 0$$

Let Φ be the set of all flows. We have:

$$\varphi_1,\ \varphi_2 \in \Phi \Rightarrow \partial(\varphi_1+\varphi_2) = \partial\varphi_1+\partial\varphi_2 = 0 \Rightarrow \varphi_1+\varphi_2 \in \Phi$$
$$\varphi\in\Phi,\ n\in\mathbf{Z} \Rightarrow \partial n\varphi = n\,\partial\varphi = 0 \Rightarrow n\varphi \in \Phi$$

Thus Φ is closed under addition and multiplication by an integer and is itself a module on $\mathbf{Z}$, a submodule of $\mathbf{Z}^U$.

Flows can be defined in an equivalent fashion, but more graphically, by saying that on each vertex of G the sum of the numbers on the 'entering arcs' is equal to the sum of the numbers on the 'leaving arcs'. This is known as Kirchhoff's law for electric currents.

Example of a flow: $\varphi = 4a+b+3c-d+2e+2f-h$

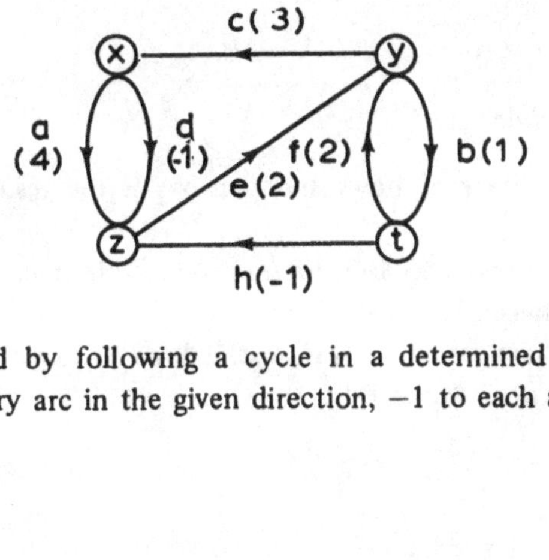

Fig. 32

A flow is obtained by following a cycle in a determined directions and by assigning +1 to every arc in the given direction, −1 to each arc in the opposite direction. $\varphi_1 = c+d-h+f$ or $\varphi_2 = a-d$ are such flows in the graph above.

3.2.2 Tensions in a graph

We will call a *potential* an element of $\mathbf{Z}^X$. The *tension* corresponding to the potential π is the element of $\mathbf{Z}^U$ which assigns to each arc u the integer:

$$\pi[T(u)] - \pi[I(u)]$$

Thus, the potential $2x-y$ gives the tension $-2a+b+4c-2d-e-f$ in our graph:

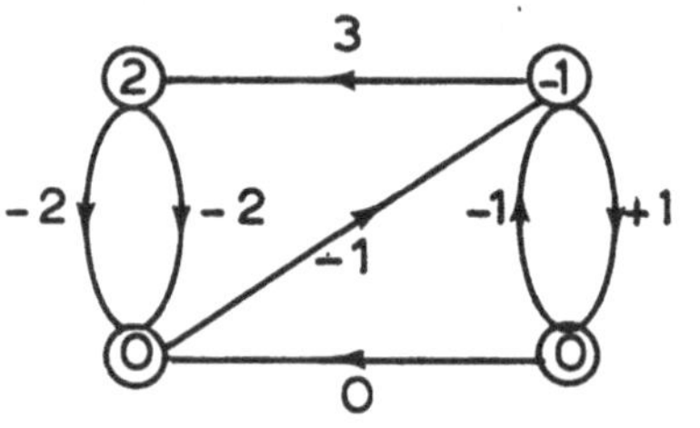

Fig. 33

As was the case for flows, tensions form a submodule of $\mathbf{Z}^U$ which we designate by Ω.

– The sum of two tensions is a tension given by the sum of their potentials.
– The tension ω corresponding to the potential π, multiplied by an integer n is the tension given by the potential $n\pi$.

Let y be a subset of X. To this subset can be associated the potential that makes +1 correspond to every vertex Y, 0 to every other. The corresponding tension is written in the form of the sum of the arcs 'entering Y' less the sum of the arcs 'leaving Y'.

Thus, to the potential $\pi = x+z$ corresponds the tension $\omega = c-e+h$ in our example.

3.2.3 Trees and modules

The importance of trees in flows and tensions is the result of the following important theorem:

1. The map which assigns to each tension $\omega \in \Omega$ its trace on the tree A is an isomorphism between Ω and $\mathbf{Z}^A$.
2. The map which assigns to each flow $\varphi \in \Phi$ its trace on the co-tree A' is an isomorphism between Φ and $\mathbf{Z}^{A'}$.

Example: Let us take the tree $\{c, e, h\}$ in the graph of Fig. 32 and 33. The tension of Fig. 33 has as its trace on A the element $4c-e$ of $\mathbf{Z}^A$. While the flow in Fig. 32 has as its tree on A' the element $4a+b-d+2f$ of $\mathbf{Z}^{A'}$.

To define a tension (or alternatively a flow) is thus to list $n-1$ (or $m-n+1$ if it is a flow) numbers, and not m which is all we need when a specific tree has been chosen. The operations of modules (addition and multiplication by an integer) will be made on these reduced lists.

Let us show by an example the exact role of a tree. In Fig. 34 below the arcs of the tree $\{c, e, h\} = A$ are stressed.

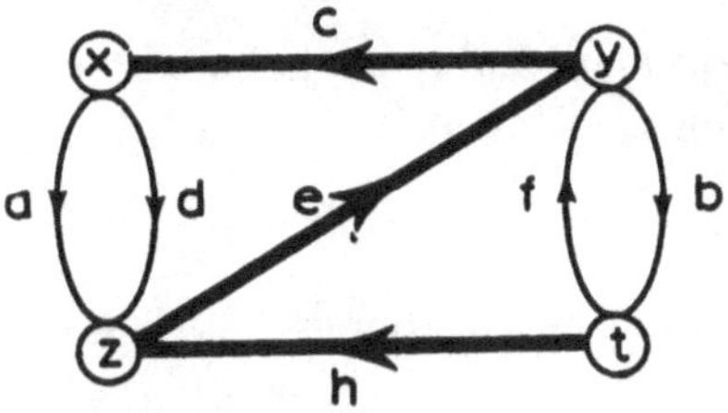

Fig. 34

The arc a determines with a subset of the tree, the cycle $\{a, e, c\}$ from which, as we have seen in 3.2.1 we obtain the flow

$$\varphi_a = a+e+c$$

Similarly

$$b \leftrightarrow \varphi_b = b+e+h$$
$$d \leftrightarrow \varphi_d = d+e+c$$
$$f \leftrightarrow \varphi_f = f-e-h$$

These flows are independent. For example φ_a is the only one to assume a nonzero value on the arc a: thus it cannot be a linear combination of the others. It follows from the theorem above that $\varphi_a, \varphi_b, \varphi_d, \varphi_f$ all generate flows. In fact, let us take a flow φ such that

$$\varphi(a) = 4 \qquad \varphi(b) = 1 \qquad \varphi(d) = -1 \qquad \varphi(f) = 2$$

The theorem states that this flow is unique. Thus we have:

$$\begin{aligned}\varphi &= 4\varphi_a+\varphi_b-\varphi_d+2\varphi_f \\ &= 4a+b+3c-d+2e+2f-h\end{aligned}$$

This is the flow shown in Fig. 34.

The deletion of an edge of a tree destroys its connectivity. Thus the deletion of c divides the partial graph $G(A)$ into two subgraphs, the set of vertices $\{x\}$ on the one hand, and $\{y, z, t\}$ on the other. As we *have already done to this* partition of X, we can associate the tension

$$\omega_c = c-a-d$$

and

$$e \leftrightarrow \omega_e = e-a-b-d+f$$
$$h \leftrightarrow \omega_h = h-b+f$$

These tensions are independent and generate the whole of Ω. Indeed, let us take as an example the tension ω such that

$$\omega(c) = 3 \qquad \omega(e) = -1 \qquad \omega(h) = 0$$

The only tension that satisfies this relation is:

$$\omega = 3\omega_c - \omega_e$$
$$= -2a+b+3c-2d-e-f$$

What we have just shown for a simple example is a general result. The modules Φ and Ω are furthermore othogonal. Therefore we have:

$$\forall \varphi \in \Phi, \qquad \forall \omega \in \Omega$$
$$\varphi \cdot \omega = \sum_{u \in U} \varphi(u) \cdot \omega(u) = 0$$

This orthogonality is very easily demonstrated by first taking the simple flows and tensions, we have just considered, and generating all of the others.

3.3 Cycles and cocycles

3.3.1 Let us replace **Z** by the field **2** of 'integers modulo 2', derived from **Z** by setting 1+1 = 0, whence −1 = +1, giving us the mapping:

$n \to 0$ if n is even; and
$n \to 1$ if n is odd.

The addition and multiplication tables of **2** are:

+	0	1
0	0	1
1	1	0

–	0	1
0	0	0
1	0	1

Let p be an element of 2^U; p is a mapping of U into 2. If $p(a) = p(d) = 1$ and $p(u) = 0$, $\forall u \in \{a, d\}$, we can write $p = a+d$. We can make the subset $\{a, d\}$ of U, correspond to p. This correspondance is a bijection and in what follows we will identify elements of 2^U and subsets of U, which will premit us to use set symbols.

2^U is a vector space with addition defined by:

$$\forall u \in U, \qquad u+u = (1+1)\,u = 0$$

This is the same sum given in section 3.1.1 on boundaries. We will, in fact, find this first type of boundary, thus completing its study. Therefore, if

$$p = a+d \text{ and } q = a+b+h$$
$$p+q = a+d+a+b+h = b+d+h$$

which is the symmetric difference of sets. More generally if $p_1, \ldots, p_r$ are elements of 2^U (subsets of U) $p = \sum_{i=1}^{r} p_i$ is the sum of elements p_i present an odd number of times.

Scalar multiplication for this set is trivial (we can only multiply by 0 or 1). A peculiarity of vector spaces on 2 is that every subgroup for addition is a vector subspace.

3.3.2 Cycles and cocycles

In this section we shall return to some of our results about flows and tensions: a part of what has already been said thus remains valid for this discussion. We shall also consider certain notions already introduced.

The boundary mapping $\partial u = T(u) - I(u)$ becomes with our rule $-x = +x$:

$$\partial u = T(u) + I(u)$$

We thus 'lose' the orientation of the arcs since $I(u)$ and $T(u)$ play the same role.

∂ is a morphism of 2^U into 2^X. We shall define a cycle as any set of arcs φ such that $\partial\varphi = 0$.

Therefore, any vertex of a graph G is attached to an even number of arcs of φ. The cycles form a subgroup of 2^U: the sum of two cycles is a cycle.

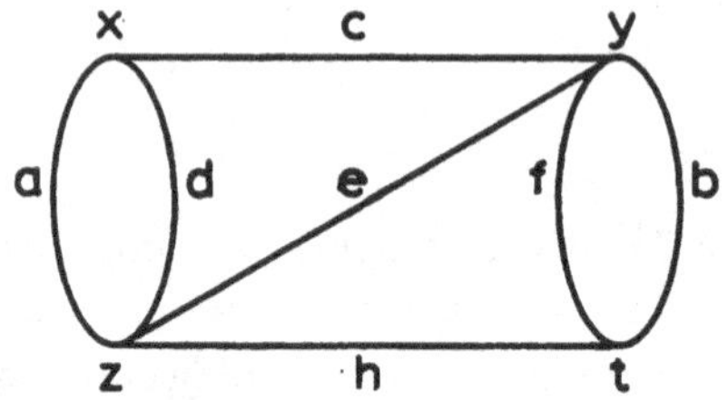

Fig. 35

$\varphi_1 = c+d+e \quad \varphi_2 = e+h+f \quad \varphi_1+\varphi_2 = c+d+h+f$

The notion of tension given in section 3.2.2 corresponds here to that of a cocycle. The cocycle $\omega(Y)$ is the sum of the arcs which have one extremity in Y and the other in $X-Y$. We thus have:

$$\omega(Y) = \omega(X-Y)$$

and

$$\omega(Y)+\omega(Y') = \omega(Y+Y')$$

where the + sign represents both types of addition defined above. In particular, we have for any cocycle:

$$\omega(Y) = \sum_{x \in Y} \omega(x)$$

The set, Ω, of cocycles is a subgroup of 2^U. As there are elementary cycles so there are elementary cocycles, those that do not include other cocycles.

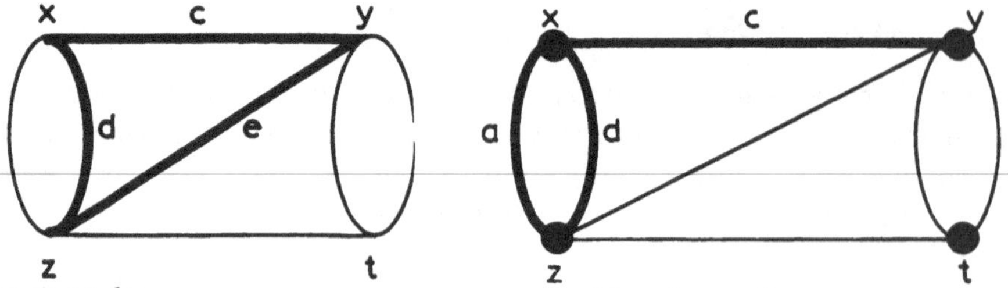

cycle $c+d+e$
$\partial\,(c+d+e) = x+y+x+t+y+t = 0$

cocycle $\omega(x) = \omega(y+z+t)$
$= a+d+c$

Fig. 36. *Cycle and co-cycle*

Example: In Fig. 35, the cocycle $\omega_1 = c+e+h$ is elementary ($\omega_1 = \omega(x+z)$). The cocycle $\omega_2 = \omega(x+t) = a+b+c+d+f+h$ is not elementary. It is the sum of the disjoint elementary cocycles $\omega(x) = a+c+d$ and $\omega(t) = b+f+h$.

These cycles and cocycles are the 'base of oddness' of flows and tensions: to any flows $\vec{\varphi}$ corresponds the cycle φ, which is the sum of the arcs u, such that $\vec{\varphi}(u)$ is odd. To every tension $\vec{\omega}$ corresponds the cocycle $\vec{\omega}$, the sum of the arcs U such that $\vec{\omega}(U)$ is odd.
Therefore the flow in Fig. 32 gives the cycle:

$$c+d+h+b.$$

The potential shown in Fig. 33 gives the cocycle:

$$\omega(y) = c+e+f+b.$$

3.3.3 Let A be a tree of G, and $A' = U-A$ the corresponding cotree. The theorem of section 3.2.3 can be transformed as follows:
- The map Ω into 2^A defined by $\omega \to \omega \cap A$ is an isomorphism.
- The map Φ into $2^{A'}$ defined by $\varphi \to \varphi \cap A'$ is an isomorphism.

There is a bijection between the cocycles and their intersection with A, and furthermore we have $(\omega+\omega') \cap A = (\omega \cap A) + (\omega' \cap A)$. Similarly for the cycles and A'.

This can be clarified by an example: the edges of the tree $A = \{c, e, h\}$ are stressed.

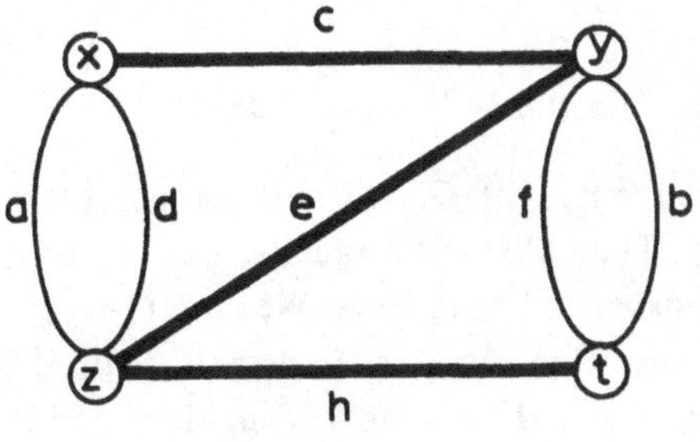

Fig. 37

As in section 2.3.2:

$$a \to \varphi_a = a+c+e$$

$$b \to \varphi_b = b+e+h$$

$$d \to \varphi_d = d+c+e$$

$$f \to \varphi_f = f+e+h$$

From A', for example, $b+d$ corresponds to a cycle:

$$\varphi_a + \varphi_d = b+c+d+h$$

All of the cycles can be obtained in this manner. Therefore there are as many cycles as subsets of A', that is 2^{m-n+1} including the 'empty cycle'.

Similarly

$$c \rightarrow \omega_c = c+a+d$$

$$e \rightarrow \omega_e = e+a+d+f+b$$

$$h \rightarrow \omega_h = h+b+f$$

And there are as many cocycles as there are subsets of A, that is 2^{n-1}, including the 'empty cocycles', which is no other than $\omega(\phi) = \omega(X)$. From these results, two new definitions of trees can be obtained.

Tenth definition of trees
Let G *be a graph. A tree of* G *is a subset of its set of arcs such that:*
1. *it intersects every non-empty cocycle of* G*; and*
2. *it is minimal with respect to this property.*

A graph-tree is a graph (X, U, I, T) *for which U is a tree.*

Eleventh definition of trees
A tree is a connected graph such that every subset of its lines is a cocycle.

3.4 The minimum tree of a graph: 'minimax' paths

Suppose we are given edges of G, without loops, totally ordered and non-directed (orientated). Then there is a specific tree A_o of G called the *minimal tree* which can be constructed as follows: We have $U = \{u_1, \ldots, u_m\}$ and $u_1 < u_2 < \ldots < u_m$. Then $u_1 \in A_o$, and $u_2 \in A_o \Leftrightarrow u_1 + u_2 \notin \Phi$ the set of cycles of G. And in general $u_r \in A_o$ if, and only if, u_r does not form a cycle with the edges already included in A_o.

Finally, we obtain a set of edges without a maximal cycle, because any edge not included gives a cycle with A_o, and hence A_o is a tree of G; Moreover, A_o has the following properties:

1. the smallest edge of any cocycle is in A_o; and
2. the greatest edge of any cycle is in $U - A_o = A'_o$.

A'_o is the *maximal cotree* of the graph G. Let us take the alphabetic order for the edges of the graph of Fig. 37: the minimal tree is $A_O = \{a, b, c\}$ the maximal cotree is $A'_O = \{d, e, f, h\}$.

An example of application. The vertices of G are the objects between which we have defined a distance index: the higher the index, the 'farther' the objects are apart. We wish to go from x to y by a path such that the greatest 'distance' on the path will be the smallest possible (we wish to minimize the greatest 'distance'): this is minimax. Thus we will take the path contained in the minimal tree of the graph (with the order of the edges furnished by the index). Another path determines, in fact, at least one cycle with this one, and the longest edge of this cycle is not in the minimal tree.

3.5 The transport problem

The transport problem is the following: we are given a set of sources or distributors $S = \{s_1, \ldots, s_i, \ldots, s_p\}$; a set of wells or clients $T = \{t_1, \ldots, t_j, \ldots, t_q\}$ a set of communications between the sources and the wells. $u_{ij} \in U$ connects the source s_i to the well t_j. Each source s_i has available a quantity ξ_i of material for which each well requires a quantity η_j.

Symbolically we have:

$$\sum_{i=1}^{p} \xi_i = \sum_{j=1}^{q} \eta_j$$

The problem is to optimize the traffic, taking into account the limited capacities r_{ij} and the cost of transport c_{ij} per item of material on the connection u_{ij}

We can express this with a graph, to which we can add one principal source s_O and a principal well to, as well as a return arc u_O, as is shown in the following diagram:

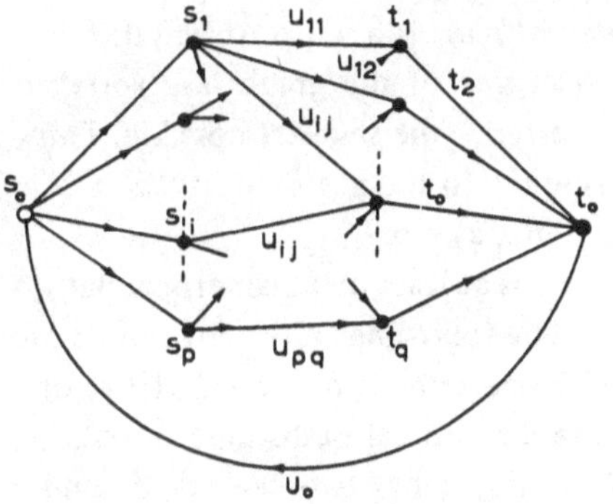

Fig. 38

The problem is thus to find the optimal flow on this graph. Another aspect of the role of trees is that in which the capacities r_{ij} are infinite. Let us consider a simple example:

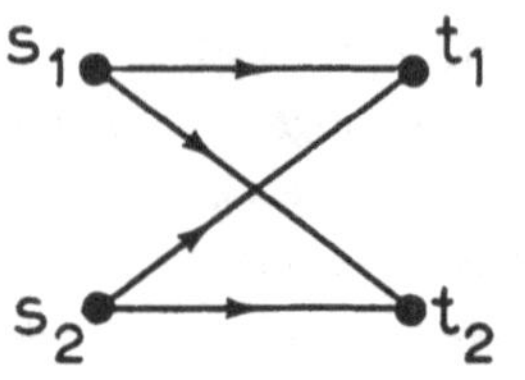

Fig. 39

Let us suppose that the most economical cost of transport is u_{11} : it would be advantageous to have the maximum amount of marchandize go by this arc. That is, all that t_1 requires and all is furnished by s_1 will use that path.

In any case, one of the axes u_{12} or u_{21} will remain empty. The transportation will not be affected on cycles composed of the edges of the graphs but rather on those of a tree. This can be generalized to having any number of vertices.

Exercise. Let G be a finite and connected graph, and X its set of vertices. The length of a word is defined as the number of letters of this word. The distance $d(x, y)$ between two vertices x and y is the smallest length of the word connecting x and y (the length of the 'shortest path' between x and y). $d(x, y)$ will be such that:

1. for every x of X, $d(x, x) = 0$;
2. for every pair of vertices (x, y), $d(x, y) \geqslant 0$; and
3. for every triplet of vertices (x, y, z), $d(x, z) \leqslant d(x, y) + d(y, z)$ (triangle inequality).

These properties make d a distance in the mathematical sense of the term. A diameter δ of the graph is the greatest of the distances between any two vertices of the graph. A *diametrial path* is a word of length δ connecting two vertices x and y: $d(x, y) = \delta$. A *center* of the graph is a vertex c such that its greatest distance to any other vertex is the smallest possible. This distance ρ is the radius of the graph. It corresponds to the notion of *radial path* of length ρ, connecting c to a vertex z such that $d(c, z) = \rho$.

A finite graph always has at least one diametrical path, at least one radial path and at least one center. The following statement can be proven:

Let A be a tree of diameter δ. If δ is even, A has only one center c and its radius is $\rho = \frac{1}{2}\delta$. All the diametrical paths pass through c, and are the union of two radial paths. If δ is odd, A has two centers, c_1 and c_2, and $\rho = \frac{1}{2}(\delta+1)$. All the diametrical paths go through c_1 and c_2 by a 'central edge (c_1, c_2). They are the union of two radial paths one for each center.

To prove this we begin by considering a diametrical path of the graph and on this path find the location of each central point. Throughout we make use of the fact that the elementary path joining two vertices of a tree is unique.

Exercise: Let D be a straight line. The interval between two points a and b is the set of points located between a and b on D, a and b can be included or excluded. For every set $\{a, b, c\}$ of three distinct points on the line there is a point common to all intervals that contain al least two of the three points a, b, c. There is a point b, for example, which is between the other two. Every interval containing a and c also contains b. We wish to show an analagous property for the words of a tree. We are given a tree. For any set $\{a, b, c\}$ of three distinct points of the tree, there is within it a point through which pass all the words of the tree (in the sense of section 1.2) which pass through two of the three points a, b, c (by saying that a word passes through a point if this point is to the left or the right of at least one letter of the word).

We can begin by considering the case, analagous to that of the straight line, where the shortest word joining the two others passes through one of the three points a, b, c. Alternatively, we could turn our attention to the shortest words between a and b, b and c, c and a, and show the existence of a 'crossroad'. The desired result is equivalent to the following.

If each pair of words of a tree pass through a common point, there is a point of the tree common to all words. This is Helly's property, for which the tree constitutes a generalization of line segments.

Bibliography

1. Beckenbach, E. F., 'Network flow problems', in *Applied combinatorial mathematics.* New York, John Wiley, 1964, pp. 348–365.
2. Bertier, P., and Roy, B., 'Une procédure de résolution pour une classe de problèmes pouvant posséder un caractère combinatoire' *Bulletin International de Calcul de Rome,* 1965.
3. Berge, C., *Graphes et hypergraphes.* Paris, Dunod, 1970 (see chap. 3 'Arbres et arborescences', pp. 22–50).
4. – and Ghouila-Houri, A., *Programmes, jeux et réseaux de transport.* Paris, Dunod, 1962 (see 'Arbres et coarbres', pp. 131–142).

5. Busacker, R. G., and Saaty, T. L., *Finite graphs and networks. An introduction with applications.* New York, McGraw-Hill, 1965 (pp. 15–18, 29–30, 58–64).
6. Flament, C., *Théorie des graphes et structures sociales.* Paris-La Haye, Mouton Gauthier-Villars (Mathématiques et sciences de l'homme, vol. 2), 1965. (see 'Longueurs et écarts dans un graphe', pp. 39–47. and 'Les réseau de communications', pp. 67–116).
7. Ford, L. R. Jr., and Fulkerson, D. R., *Flots dans les graphes,* trans. by J.-C. Arinal. Paris, Gauthier-Villars, 1967 (pp. 192–213).
8. Frank, H., and Frisch, I. T., *Communication, transmission and transportation networks.* New York, Addison-Wesley, 1971.
9. Gross, M., and Lentin, A., *Notions sur les grammaires formelles.* Paris, Gauthier-Villars, 1967 (see 'Langages de Dyck', pp. 159–163).
10. Guilbaud, G. Th., 'Greffes et élagages d'arbres', *Mathématiques et Sciences Humaines,* n° 21, 1968 (see 'L'algorithme de l'élagueur').
11. –, Massé, P., and Henan, R., *Stratégies et décisions économiques.* Etudes théoriques et applications aux entreprises. Cours et Conférences. Paris, C.N.R.S., Centre d'économetrie, 1951–1953, 1954.
12. Ginsburg, S., *The mathematical theory of context-free languages.* New York, McGraw-Hill, 1966 (see 'Generation Trees', pp. 23–29).
13. Iverson, K. E., *A programming language.* New York, John Wiley, 1962 (see 'Ordered trees', pp. 45–62 and 'Representation of trees', pp. 121–128).
14. Kaufmann, A., 'Méthode de séparation et évaluations progressives [Branch and bound method]', in *Introduction à la combinatorique.* Paris, Dunod, 1968 (pp. 377–407).
15. Knuth, D. E., *The art of computer programming.* Cambridge, Addison-Wesley, 1968 (pp. 305–399).
16. Kruskal, J. B. Jr., 'On the shortest spanning of a graph and the traveling salesman problem', *Proc. Amer. Math. Soc.*, 1956/7, pp. 48–50.
17. Little, J. D. C., *et al.*, 'An algorithm for the traveling salesman problem', *J.O.R.S.A.*, 1963 (11), pp. 972–989.
18. Liu, C. L., *Introduction to combinatorial mathematics.* New York, McGraw-Hill, 1968, (see 'Trees, circuits and cut-sets', pp. 185–201).
19. McMahon, *Combinatory analysis,* 2nd ed. New York, Chelsea Publishing Co., 1960 (see 'Correspondence with the theory of trees', pp. 181–182).
20. Ore, O., 'Theory of graphs', *Amer. Math. Soc. Colloq. Publ.* Providence, The American Mathematical Society, 1962, vol. 38 (see 'Trees', pp. 58–77).
21. – *Les graphes et leur applications.* Paris, Dunod (see 'Arbres', pp. 34–44).

22. Raiffa, H., *Decision analysis.* Cambridge, Addison-Wesley, 1968 (see especially 'The decision flow diagram', pp. 10–13).
23. Rosenstiehl, P. (in collaboration with Bermond, J. C. et Moniez, F.), *Graphes, leurs vecteurs et leurs mots.* Cours de l'École Polytechnique. Paris, 1973.
24. Roy, B., *Algèbre moderne et théorie des graphes.* Paris, Dunod, vol. I, 1969 (see 'Graphes sans cycles', pp. 334–347 and 'Exercices', pp. 339–404); vol. II, 1970 (see chap. 10 'Procédures d'exploration', pp. 653–744).
25. Veblen, O., 'Analysis situs', *Amer. Math. Soc. Colloq. Publ.* 1931, vol. 5/2, 2nd ed. (see 'Trees', pp. 18–19).

2. ordering and classification

by B. Monjardet

Classifications and hierarchies are common in society: individuals are classified at school, in the army and in civilian or professional life. The criteria of classification can be age, height, 'average grade', family situation, professional rank, income, etc. The individual himself has preferences when, for example, he establishes an ordering on the set of goods he will consume. At the collective level, an enterprise has multiple problems of choice: investments, commercial strategies, publicity methods . . . For decisions to be made, it is necessary to establish an ordering within the set of possible choices. On a more technical level, an entreprise, an administration, or even an individual has problems related to the classification of documentation. In short, a multitude of facts, social or individual, public or private, have a classification component.

What are the characteristics common to all classification situations? We begin with a set of objects to order; the very operation of ordering introduces a relationship between these objects since two objects are either placed together, or one before the other. In both of these cases, if the object A is classed before (or with) object B, and if object B is placed before (or with) object C, object A has been classed before (or with) object C. In the language of mathematics, we say that the ordering relation is transitive. The mathematical study of ordering and classification is thus the study of transitive relations. These relations are called quasiordering relations if they satisfy, in addition, the property of transitivity and the property of reflexivity, which though minor in themselves, are useful from the point of view of mathematical technique.

Section 1 of this chapter will be thus devoted to the study of the quasi-ordering relations of a set. We have just emphasized the property that is common to ordering: transitivity, but there are also some important differences. If A is ordered *with* B, while B is ordered *with* A, then there is a symmetry in the relation 'ordered with'. On the other hand, if A is ordered *before* B, while B is *not* ordered before *C,* then we have the converse of symmetry, and we say that the relation 'ordered before' is antisymmetric. Thus, we shall distinguish between symmetrical quasiordering – also called equivalence relations – and antisymmetrical quasiordering also called ordering relations. The former will be studied in section 2 where we will show their equivalence with the 'partitions' of a set. The latter will be studied in section 3 where we shall define certain orderings of particular importance such as lattices and trees. What we learn from these two sections will permit us, in section 4, to return to the case of any quasiordering and show that they are always 'decomposable' into an equivalence and an ordering. Finally, in section 5 some applications of the notions already presented will be given.

It should be pointed out that the mathematical concepts introduced in this article are neither numerous nor complicated. They essentially involve such relations as binary, quasiordering, equivalence, partitions, orderings, trees and lattices. Nevertheless, they suffice to formalize the non-negligible aspects of a host of daily situations. The mathematical model thus introduced is hardly onerous and is very useful. First, because while some of those properties that follow the model are intuitive and correspond to our experience, there are others that are contrary to 'common sense' and require the aid of a 'calculus'. Moreover, because the language of mathematics is precise it often leads to a clarification useful in the investigatory methods of the social sciences. As for the inherent limitation of mathematical models, the specialists in the social sciences will be even more conscious of them when he has a better understanding of the mathematics to which this article is to introduce him.

1. Quasiorderings

A quasiordering on a set E *is a binary relation on the set* E *which is transitive and reflexive.*

We shall first make explicit the notions of binary relations of transivity and reflexivity. Then we shall give some examples of quasiordering, and finish with a typology of quasiorderings.

1.1 Binary relation

Let E be a set. Recall that the cartesian product of the set E on itself is the set of ordered pairs (x, y), where $x \in E$ and $y \in E$. We shall write this set as $E \times E$ or

E^2. It can be represented by a two-way table, where each box represents a pair. For example, if $E = \{a, b, c, d\}$, E^2 can be represented as follows:

→	*a*	*b*	*c*	*d*
a	(*a*, *a*)			
b			(*b*, *c*)	
c		(*c*, *b*)		
d				

A binary relation *R* on a set *E* is a subset of the *cartesian product* E^2.

$$R \subset E^2$$

If the pair (x, y) belong to the relation R; $(x, y) \in R$, and we say that x is in relation with y, which can also be written: $x\,R\,y$.

We can consider all the elements with which a particular element x is in relation. Thus we obtain a subset of E, which may be empty, and write it $R\,x$. One of the ways in which we obtain a relation is precisely that of forming the subsets $R\,x$ for all the elements of E, as in the following example.

Example 1: We shall examine the forces of attraction or repulsion in the interior of a group of eight students: Armelle, Bernard, Cornelia, Dominique, Evariste, Fougère, Gaspard and Helen. Among the possible experiments is the following: we ask each student to choose the friends with whom he would like to study.

We thus obtain a table that defines a relation R in the set $E = \{a, b, c, d, e, f, g, h\}$ of eight students:

a	*b*, *d*, *g*
b	*a*, *g*
c	*f*
d	*a*, *b*
e	*b*, *c*, *d*, *g*
f	*c*, *g*
g	∅
h	*g*

$R\,a = \{b, d, g\}$

or

$a\,R\,b$, $a\,R\,d$ and $a\,R\,g$

This relation can be represented graphically in several ways – by a cartesian graph, for example. On the two-way table representing the cartesian product E^2, we can put a cross in the boxes representing the pairs of relations. Or, this relation can be represented by the network graph where each element of E is represented by a point, and the pair (x, y) belonging to the relation is represented by a directed arc going from the point representing x to the point representing y. Thus, the preceding relation can be represented as follows:

	a	*b*	*c*	*d*	*e*	*f*	*g*	*h*
a		×		×			×	
b	×						×	
c						×		
d	×	×						
e		×	×	×			×	
f			×				×	
g								
h							×	

Cartesien graph

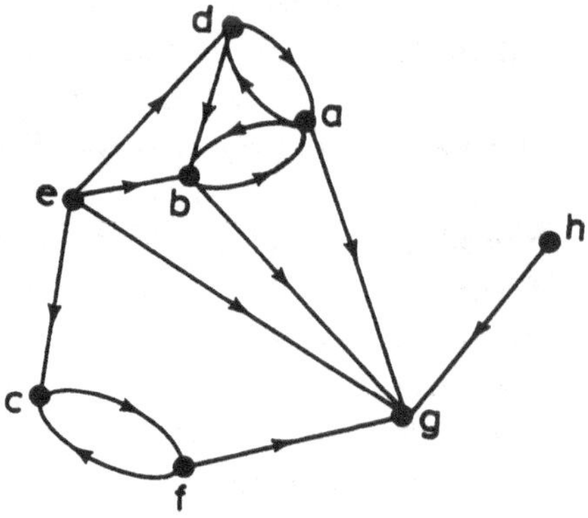

Network graph of R

Fig. 1

Note: We have defined a binary relation. We can define ternary, . . ., *n*-ary. In what is to follow, we shall restrict ourselves to binary relations which we will call simply, relations.

1.2 Transitive relation

A relation (binary) R on a set E is said to be *transitive* if each time that x is in relation with y, and y is in relation with z, x is in relation with z. This can be written symbolically:

$$\forall x \in E, \quad \forall y \in E, \quad \forall z \in E, \quad [(x\ R\ y \quad \text{and} \quad y\ R\ z) \Rightarrow (x\ R\ z)]$$

($\forall$ is read 'for all' and $\in$ is read 'belongs to'; and we shall 'decode' with the conventions given in the above expression.) On the network of a relation, we recognize transitivity if we can obtain the configuration:

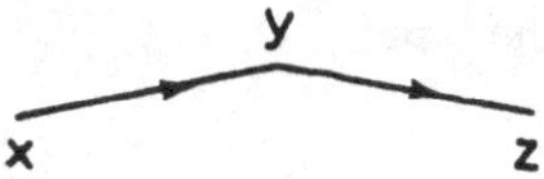

Fig. 2a

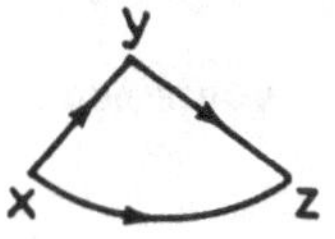

Fig. 2b

completed by an arc from x to z. The relation of this example is not transitive because we have $b\ R\ a$ and $a\ R\ d$ without having $b\ R\ d$, for example.

1.3 Reflexive relation

A (binary) relation R on a set E is said to be reflexive if every element is in relation with itself, which is written:

$$\forall x \in E, \quad x\ R\ x$$

This again means that all of the pairs (x, x) belong to the relation. Thus, in the cartesian graph of Figure 1, all of the boxes on the diagonal have a cross. In the graph of the network of R there is a loop at each vertex:

Fig. 3

The relation of Example 1 is not reflexive, but it is easy to make it reflexive by changing the definition very slightly (how?); this illustrates that reflexivity is a minor property.

1.4 Quasiorderings: examples

Recall that a quasiordering R on a set E is a (binary) relation on E which is transitive and reflexive. If for three elements x, y and z of E we have:

$x\,R\,y, \quad y\,R\,z, \quad z\,R\,x$

We say that these elements form a cycle of order 3. But according to the transitivity of quasiordering we have also:

$x\,R\,z, \quad z\,R\,y, \quad y\,R\,x$

In other words, in the network of quasiordering, a configuration of the type of cycle of order 3:

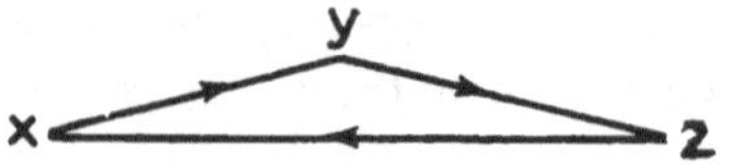

Fig. 4a

cannot exist unless it is contained in a configuration of the type:

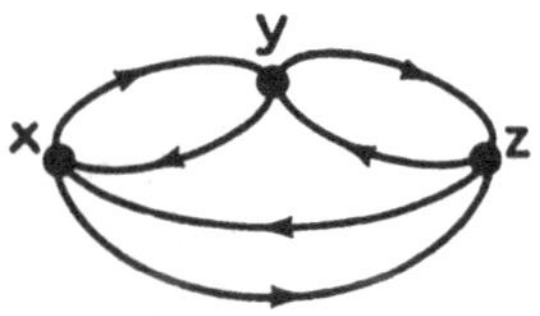

Fig. 4b

We have the same result for a cycle of n-order.

A set on which a quasiordering relation has been defined is called a *quasi-ordered set.* Such sets are common. For example, if E is a set of economic goods, the preference relation between these goods for an individual is a quasiordering relation.

If $E = P(x)$ is the set of the subsets of a set X, we define a quasiordering relation on E by writing for $A \subset X$, $B \subset X$

$$A\,R\,B \Leftrightarrow |A| \leqq |B|$$

(where $|A|$ is the number of elements, or the cardinal of the subset A).
We will give other examples later on.

Example 2: Let us return to the set E of students in example 1. In this set the relation 'x is at most as old as y' is a quasiordering relation. The ages of the students are as follows:

a	b	c	d	e	f	g	h
19	21	19	20	21	18	20	21

and the quasiordering relation is represented by the following network:

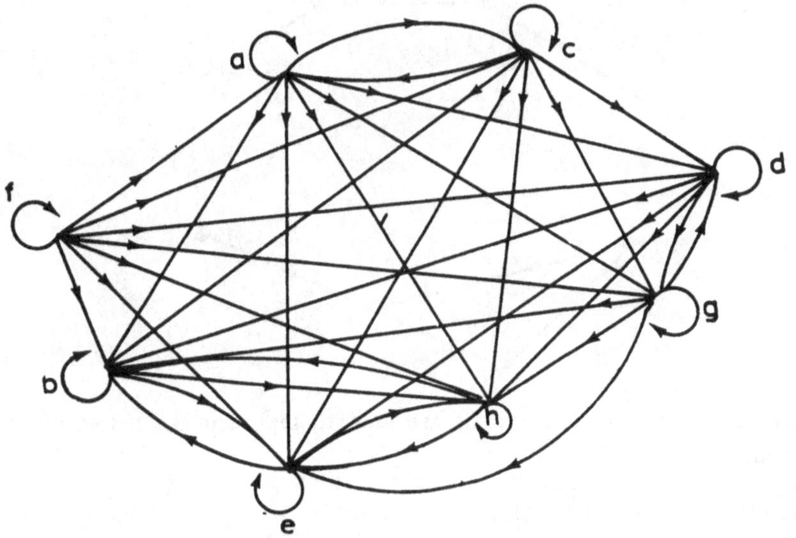

Fig. 5

This is a complicated network and to simplify its graphic representation we can adopt certain conventions:

1. A double directed arc is replaced by a single edge without an arrow:

becomes

Fig. 6

Fig. 7

2. A transitivity arc (that is, an arc obtained as the consequence of two others by transitivity) is deleted:

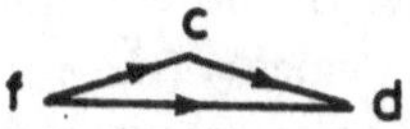

becomes

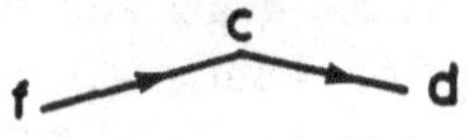

Fig. 8

Fig. 9

3. A reflexive arc (loop) is deleted:

becomes *a*.

Fig. 10

Thus, the network previously given becomes:

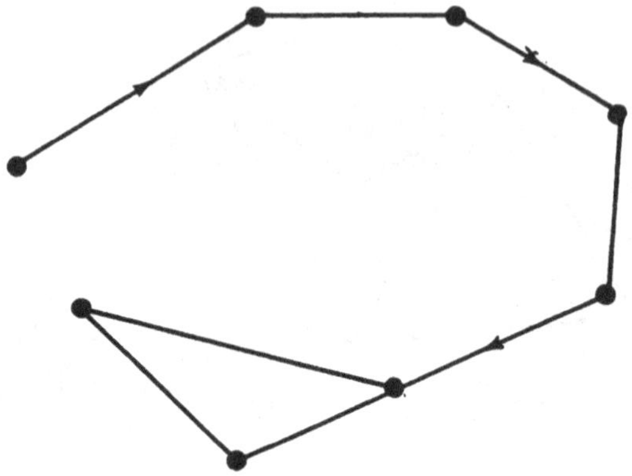

Fig. 11

In general, the preceding conventions are used to represent the networks of quasiordering.

Example 3:

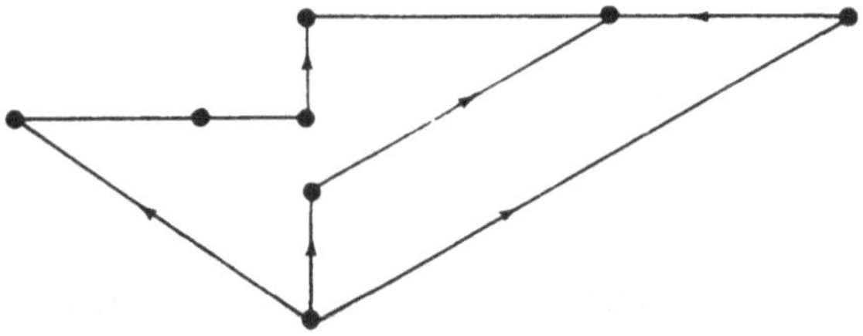

Fig. 12

Example 4:

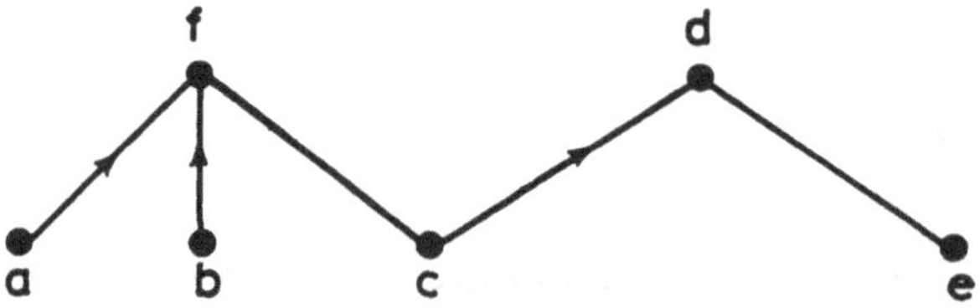

Fig. 13

Exercise 1: Before going any further, the reader should construct all of the quasiorderings on a set of first 2 and then 3 elements (there are 4 and 29 respectively). For 4, 5, 6 and 7 elements there are respectively: 355; 6,942; 209,527; and 9,535,241 distinct quasiorderings.

1.5 Typology

On a set E, even one having few elements, we can define a large number of quasiorderings (see exercise 1). To see this more clearly we shall classify these quasiorderings according to supplementary properties that can or cannot be satisfied. We shall consider three properties:

Symmetry: For any $x \in R$ and for any $y \in E$, $x \, R \, y \Leftrightarrow y \, R \, x$;

Anti-symmetry: For any $x \in E$ and for any $y \in E$, $x \neq y$ and $x \, R \, y \Rightarrow y \not R \, x$ (y (y not being in relation with x).

Totality. For any $x \in E$ and for any $y \in E$, $x \neq y$ and $x \not R \, y \Rightarrow y \, R \, x$.

A symmetric quasiordering is called an *equivalence.* According to the conventions given above, the network representing an equivalence contains only edges without arrows.

An anti-symmetric quasiordering is called an *ordering*. In a network of an ordering, two vertices are connected by at most one arc (never an edge). In the network of a total quasiordering, two vertices are connected by at least one arc.

The properties of symmetry and antisymmetry are mutually exclusive unless there exists no pair (x, y) with x different from y in the relation. In such a case, the quasiordering reduces to $D = \{(x, x) \, x \in E\}$ and is said to be a diagonal quasiordering; it is the smallest quasiordering on E. If $E = \{x, y, z\}$, this quasiordering is represented as follows (with its reflexivity arcs) and is the union of loops:

Fig. 14

On the other hand, we can consider an antisymmetric, total quasiordering. We thus obtain what is called a *total ordering.*

What can be said when a quasiordering is symmetric and total (i.e. when it is a total equivalence on a set E)? Since the quasiordering is total, any two elements are in relation; and since it is symmetric, they are in relation in both directions. Thus, any pair of elements of E is in the quasiordering and we obtain the universal relation $U = E^2$. It is the greatest quasiordering on E; it is also called the universal clique. We have represented it below for a set E having respectively one, two, three, four, and five elements (the transivity edges are shown):

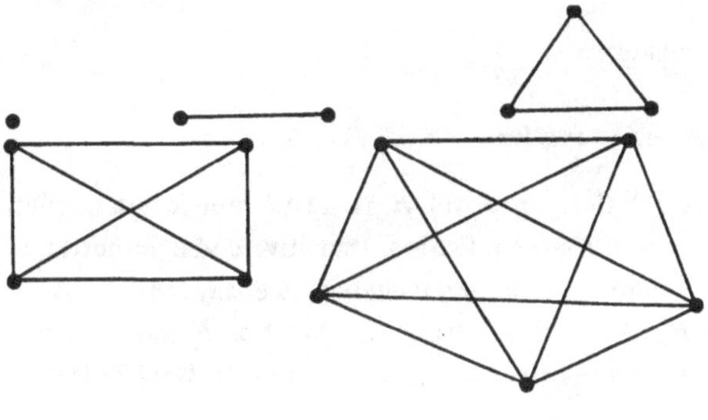

Fig. 15

The different classes of quasiorders and their relations can be summarized in the following figure:

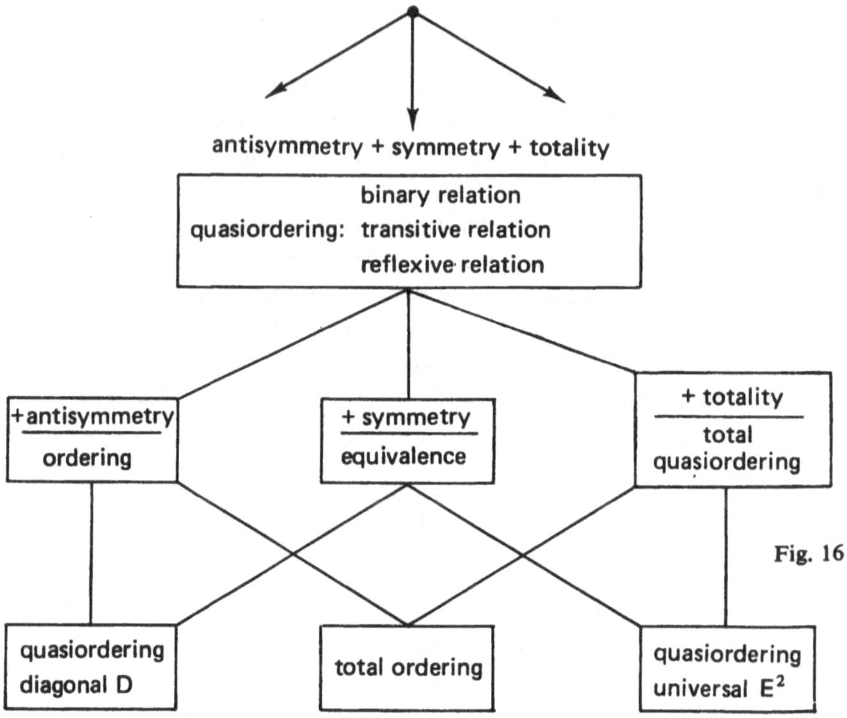

Fig. 16

Exercise 2: In the 29 quasiorderings on a set of three elements, show that there are 13 total quasiorderings among which six are total orderings and five are equivalences. We note that in this case any quasiordering is an ordering, an equivalence, or a total quasiordering. Is this always the case (see Example 1)?

2. Equivalence relations

2.1. Definitions and examples

An equivalence relation on a set E is a symmetric quasiordering on E or a (binary) relation on E that is reflexive, transitive and symmetric.

If $x\ R\ y$, where R is an equivalence, we say that x is equivalent to y (according to R). Instead of using the symbol R, an equivalence relation is often written as $\equiv$. We write $x \equiv y$ (modulo R) or $x \equiv y$ (R) if $(x, y) \in R$ and $x \not\equiv$

$y\,(R)$ if $(x, y) \notin R$. In place of the sign $\equiv$ one sometimes sees the signs $=$ or $\sim$.

Examples of equivalence relations are numerous: the equality relation in a set, the relation 'having the same parents' or the relation 'to be of the same sex' in a population, the relation $|A| = |B|$ in the set $P(X)$ of the subsets of a set X, to cite a few. In a population the relation, 'to be the brother of', is not an equivalence (why?) but it can be easily transformed into an equivalence relation (how?).

Example 5: Let us return to the set E of the eight students of Examples 1 and 2. In this set the relation 'to be of the same age' is an equivalence relation that can be represented in the form of a table and a network as follows:

	a	*c*	*b*	*e*	*h*	*f*	*d*	*g*
a	×	×						
c	×	×						
b			×	×	×			
e			×	×	×			
h			×	×	×			
f						×		
d							×	×
g							×	×

Fig. 17. *Table*

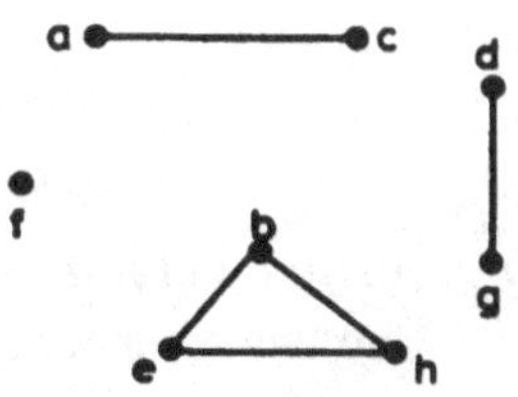

Fig. 18. *Network*

In the table the equivalent elements have been regrouped and we thus obtain squares lying the diagonal.

Exercise 3: Represent the equivalence relations on a set having respectively two, three and four elements by their networks. There will be respectively two, five and 15 networks (see Exercise 5).

Equivalence relations are extremely important; they are used whenever one wishes to classify the elements of a set. The act of classification can be seen intuitively as an act of assigning objects to different categories. The corresponding mathematical concept is that of partition. We will first present this concept and then show how, given an equivalence relation, on a set we can, in fact, define a partition of the elements of this set – that is, to classify them. For this reason, an equivalence relation is sometimes called a classification relation.

Conversely, we shall show that an equivalence relation is associated to every partition, so that these two concepts can be shown to be equivalent.

2.2 Partition

A *partition* of a set E is a set of subsets of E such that every element of E belongs to one, and only one, of these subsets.

The subsets constituting the partition is called the *partition classes.*

Let

$$\Pi = \{C_1, C_2, \ldots, C_i, \ldots, C_p\}$$
$$= \{C_i\}_{i \in I}$$

be a set of subsets of E.

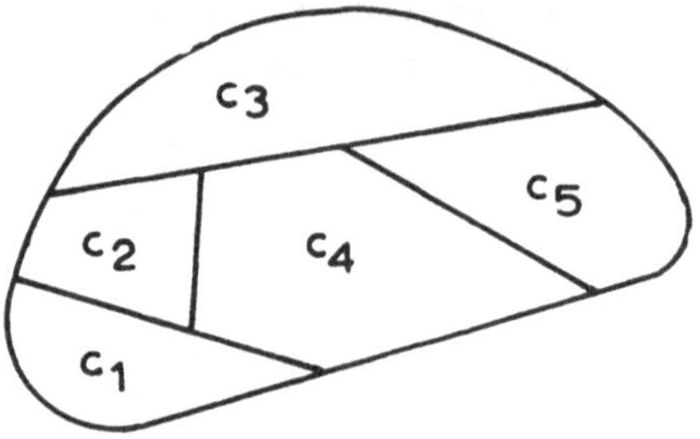

Fig. 19. *The partition into five classes*

Π is a partition of E if, and only if:

1) any two distinct subsets are disjoint:

$$C_1 \cap C_j = \emptyset \quad \forall i \in I, \quad \forall j \in I \quad \text{with} \quad i \neq j \quad \text{and}$$

2) the subsets cover the set E:

$$C_1 \cup C_2 \cup \ldots \cup C_p = \bigcup_{i \in I} C_i = E$$

Concrete examples are plentiful: the partition of a population according to age, nationality, profession; partition of the territory of a country according to administrative divisions. In France, for example, territories are partitioned into communes, cantons, departments and regions. Groups of individuals can also be partitioned according to their scores on an aptitude test, or according to their I.Q., etc. Nevertheless, it should be pointed out that in certain cases the partitions mentioned above are not partitions in the mathematical sense of the term; for example, in the case of nationality where the individual has a dual nationality, or none. In the mathematical formulation of concrete data, the concept of partition is thus not used unless it has been determined that the data satisfy properties 1 and 2 above.

If a set E has n elements, the number p of classes of a partition of E is between 1 and n. In particular, if $p = 1$, there is but one class which is thus the set E itself; this partition is sometimes called gross. Conversely, if $p = n$, there are n classes each containing a single element: the resulting partition is sometimes called trivial. A partition of E into two classes if of the form: $\{A, CA\}$ (where CA is the complementary subset of the subset A). This type of partition is found every time we need to distinguish the individuals possessing a character and those who do not in a particular population.

Example 6: $E = \{a, b, c, d, e, f, g, h\}$ is the set of the eight students already encountered.

$\Pi = \{\{a, c\}, \{d, g\}, \{b, e, h\}\}$ is thus the partition of E into four classes.

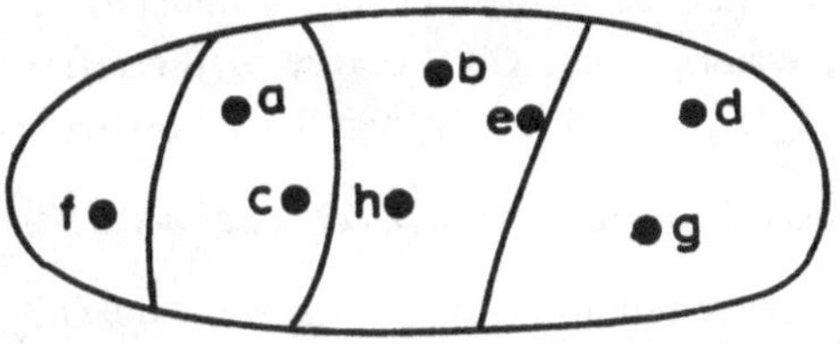

Fig. 20

Exercise 4: Enumerate all the partitions of a set E having two, three or four elements. Compare the results with exercise 3.

2.3 *Partition associated with an equivalence relation: quotient set*

The partition of example 6, based on the equivalence relation 'to be of the same age' had already been obtained in example 5. It is thus a question of the age classes of a group of students.

In general, to every equivalence relation R on a set E, is associated a partition on E: that for which the classes are equivalence classes.

Let $x \in E$, and we set $C_x = \{y \in E: x \equiv y \pmod R\}$.

C_x is the set of elements equivalent to x and is called the class of equivalences of x. Let $\Pi_R = \{C_x, x \in E\}$ then be the set of equivalence classes. Let us show that Π_R is a partition of E.

First, two equivalence classes are either disjoint or identical. Let C_x and C_y be two equivalence classes.

– If $x \equiv y \pmod R$, then we have $C_x = C_y$.

And indeed

$z \in C_x \Rightarrow z \equiv x$

$z \equiv x$ and $x \equiv y \Rightarrow z \equiv y$ (by the property of transitivity of equivalence)

$z \equiv y \Rightarrow z \in C_y$.

Thus, $C_x \subset C_y$. We can show similarly that $C_y \subset C_x$, from whence $C_x = C_y$.

– If, on the other hand, $x \not\equiv y$ (x and y nonequivalent mod R), then we have $C_x \cap C_y = \emptyset$.

Suppose, in fact, that $z \in C_x \cap C_y$, then

$$\left.\begin{matrix} z \in C_x \Rightarrow z \equiv x \\ z \in C_y \Rightarrow z \equiv y \end{matrix}\right\} \Rightarrow x \equiv y \text{ (by the symmetrical and transitivity properties)}$$

which contradicts our hypothesis. Thus, there can exist no elements common to C_x and C_y.

Secondly, the equivalence classes cover E. Indeed, any element of E is contained in its equivalence class C_x, because of the reflexivity of the equivalence relation. Thus, $\bigcup_x C_x = E$, giving us then the following result:

Proposition: Any equivalence relation on a set E defines a partition of E.

Being given an equivalence relation R on a set E we will call it the quotient set of E mod R, and write E/R as the set of equivalence classes of R. The quotient set is thus not other than the partition associated with

$$R : E/R = \Pi_R$$

Example: The set $\mathcal{P}(X)$ of the subsets of a set X having n elements and ordered by the inclusion between subsets is called the simplex S_n. We can define the relation 'to have the same number of elements' which is an equivalence relation:

$$A \equiv B \text{ (mod } R) \Leftrightarrow |A| = |B|.$$

The corresponding quotient set $\mathcal{P}(X)/R$ is composed of the 'levels' of simplexes (the set of the subsets having the same cardinal).

2.4 The equivalence relation associated with a partition

We shall now give the proposition that is the reciprocal of the one given in the preceding section:

Any partition on a set E defines an equivalence relation on E.

This proposition is 'evident' intuitively when one is familiar with the notions of relation, equivalence and partition. The formal proof is given below and the reader should study it as an exercise.

Let $\pi = \{C_1, \ldots, C_i, \ldots, C_p\}$ be a partition of E. We define a relation R_π in E by writing:

$$x\, R_\pi\, y \Leftrightarrow \exists\, i, \quad 1 \leqq i \leqq p, \quad \text{with} \quad x \in C_i \quad \text{and} \quad y \in C_i.$$

In other words, two elements are in relation if they are in the same partition class. We shall show that R_π is an equivalence relation:

1. R_π is reflexive,
   $xR_\pi x$ because π being a partition, x belongs to one of the subsets of C_i.
2. R_π is symmetric.
   $xR_\pi y \Rightarrow y\, R_\pi\, x$, which is evident from the definition of R_π.
3. R_π is transitive.

$$\left.\begin{array}{l} x\, R_\pi\, y \Leftrightarrow x \text{ and } y \in C_i \\ y\, R_\pi\, z \Leftrightarrow y \text{ and } z \in C_j \end{array}\right\} \Rightarrow y \in C_i \cap C_j$$

But since π is a partition, $C_i = C_j$. Thus we have x and $z \in C_i \Rightarrow x\, R_\pi\, z$.

The results of this and the preceding sections show that the notions of partition and equivalence relation are logically equivalent. We have thus two distinct languages to use for the classification of objects. These two languages are frequently used in practice; the language of partitions is the one we use when we speak of arranging the objects in 'boxes' (corresponding to the cells of the partition); the relational language is that employed when we speak of grouping objects of the 'same type' (the relation 'to be of the same type' being clearly an equivalence relation). We make an accord between the two languages by saying that two objects are placed in the same box if, and only if, they are of the same type (two objects are in the same class if, and only if, they are equivalent).

Exercise 5: Show that the number P_n of partitions on a set of n elements is given by the recurrence formula:

$$P_n = \binom{n-1}{0} P_{n-1} + \binom{n-1}{1} P_{n-2} + \ldots + \binom{n-1}{k} P_{n-1-k} + \ldots + \binom{n-1}{n-1} P_0$$

$\binom{n-1}{k}$ is the number of subsets of k elements in the set of $(n-1)$ elements.

We set $P_0 = 1$.

To obtain this formula, we consider a particular element x of E, and succesively count the number of partitions where x is is found among 0, 1 . . .

$n-1$ other elements of E.

Calculate P_n for $n \leqslant 6$ using the preceding formula thus obtaining the same results as in exercise 3.

2.5 Mapping and equivalence relations

We know that a mapping is defined by the following:

1. E is the initial set.
2. F is the final set.
3. A correspondence exists associating to *every* element of x in E a *unique* element y of F, called the image of x, by the mapping f (y is written $f(x)$).

Being given a mapping f of a set E into a set F, permits us to define an equivalence relation on E. Thus, let $f : E \to F$ define a binary relation R_f by setting:

$$x \, R_f \, y \Leftrightarrow f(x) = f(y)$$

Two elements of E are in relation if, and only if, they have the same image by the mapping f. Rf is, of course, reflexive and symmetric:

$$\left.\begin{array}{l} x \, R_f \, y \Rightarrow f(x) = f(y) \\ y \, R_f \, z \Rightarrow f(y) = f(z) \end{array}\right\} \Rightarrow f(x) = f(z) \Rightarrow x \, R_f \, z$$

Thus, R_f is transitive and is also an equivalence relation; we call it the equivalence induced by f into E, and the corresponding quotient set is written E/f.

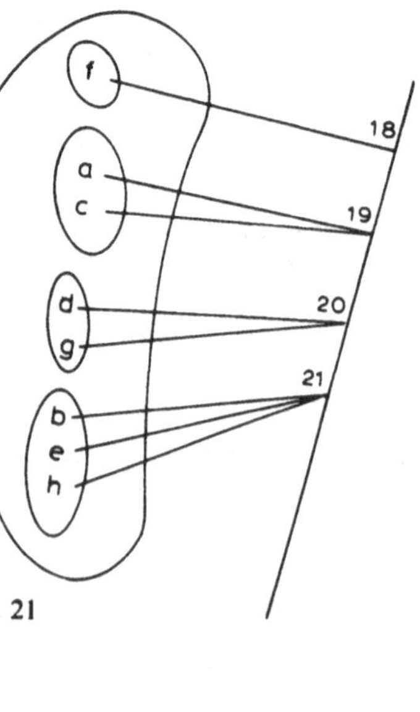

Fig. 21

Example 7: To each student in our group of students let the correspondence be made with his age. We thus define a mapping f, for which the associated equivalence is: 'to have the same age', and for which E/f is the set of age classes.

Another example is that of certain so-called primitive societies where to each individual is associated a type of marriage determined by the rules of kinship. The society if thus divided into classes sometimes called matrimonial classes (see chapter V, 'Monoids and groups').

To every mapping of E into F we associate an equivalence relation on E. Conversely, can an equivalence R on E be obtained from a mapping of E into another set? Yes. It is sufficient to consider the quotient set E/R and a mapping of E into E/R which, to every element of an equivalence class R, makes correspond an element of E/R. How many ways of proceding are there in the case of Example 7? Can you generalize?

3. Orderings

3.1 Definitions and examples

An ordering on a set E *is a quasiordering, antisymmetric on* E – *that is, a (binary) relation on* E – *which is reflexive, transitive and antisymmetric.*

An ordering R on a set E is often noted $\leqslant$, read less than or equal to; we write $x \leqslant y$ if $(x, y) \in R$; $x \nleqslant y$ if $(x, y) \notin R$. Instead of writing $x \leqslant y$ we can also write $y \geqslant x$, which is read as: y is greater than or equal to x. Finally, $x \leqslant y$ and $x \neq y$ is written $x < y$ or $y > x$.

A set for which an ordering relation has been defined is called an ordered set and is denoted by $(E, \leqslant)$.

Examples:

1. The set N of natural numbers is a set ordered by the usual relation: $3 \leqslant 5$.

2. The set $\mathcal{P}(X)$ of the subsets of a set X is an ordered set for the inclusion relation $A \subset B$. We can give several examples. Thus, if $X = E^2$ is the cartesian product of E on itself, E^2 is the set of binary relations on E. This set is thus ordered by the relation: all pairs in the relation R are in the relation S.

$$R \subset S \Leftrightarrow [\forall x \in E, \forall y \in E, (x, y) \in R \Rightarrow (x, y) \in S];$$

in this case we can also say that the relation R is compatible with the relation S.

3. The set of the letters of the alphabet is ordered by the alphabetic ordering: 'a before b'.

4. The set of words in a dictionary is ordered by the lexicographic ordering: 'amorous before amuse', for which the reader is to find the precise definition.

Exercise 6: Show that contrary to any quasiordering relation, an ordering relation cannot have a cycle of the form:

$$x_1 \leqq x_2 \leqq \ldots x_i \leqq x_{i+1} \ldots x_{n-1} \leqq x_n \quad \text{with } x_1 = x_n$$

3.1.1 Networks and covering relations

In the graphic representation of an ordering relation by its network, the arcs of reflexivity and transitivity are not drawn. We also replace the arcs by edges by agreeing to the convention by which, for example, the edges are orientated from bottom to the top of the diagram. Thus, for the set $E = \{a, b, c, d, e\}$ the ordering R, defined by

$$a < b, a < c, a < d, a < e, b < c, b < d, b < e, d < e$$

is represented by the following network:
In such a network, two vertices x and y are linked by an edge if we have $x < y$ and if there exists no element z such that $x < z < y$. We then say that the element y covers the element x and that the element x is covered by the element y. This covering relation is denoted by $<$; it is antisymmetric but not transitive.

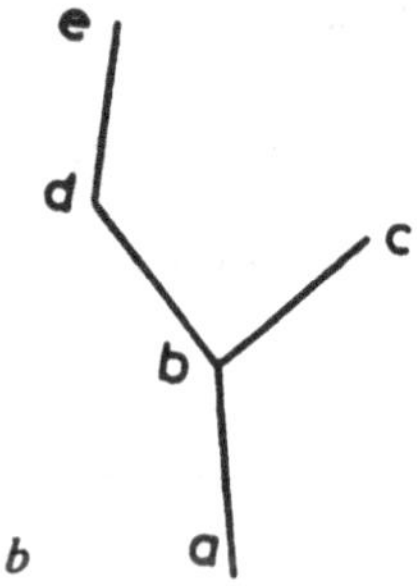

$a < b$

Fig. 22

Exercise 7: Let N be the set of positive integers or zero. We define the binary relation:

$$x\, R\, y \Leftrightarrow x \text{ divides } y.$$

Show that N is a set ordered by this relation. If a set is ordered, every subset of this set is also ordered by the 'restriction' of the ordering relation of this subset. Draw the network of the following subsets of N ordered by the relation of divisibility:

$$\{1, 2, 4, 8\} - \{1, 2, 3, 6, 8\} - \{1, 2, 3, 4, 6, 12\}$$

3.1.2 Ordering and strict ordering

Let $(E, \leqslant)$ be an ordered set. Consider the relation on E defined by $x\, R\, y \Leftrightarrow \begin{cases} x \leqq y \\ x \neq y \end{cases}$; This relation is never reflexive: $x \not R x$ for all x. It is antireflexive, antisymmetric and transitive. Such a relation is called a strict ordering and it is written $x < y$. It should be noted that this is not an ordering in the sense we have just defined since it is not reflexive.

Example 8: In our group of eight students, the relation 'to be younger than' is a strict order relation for which the network is drawn below.

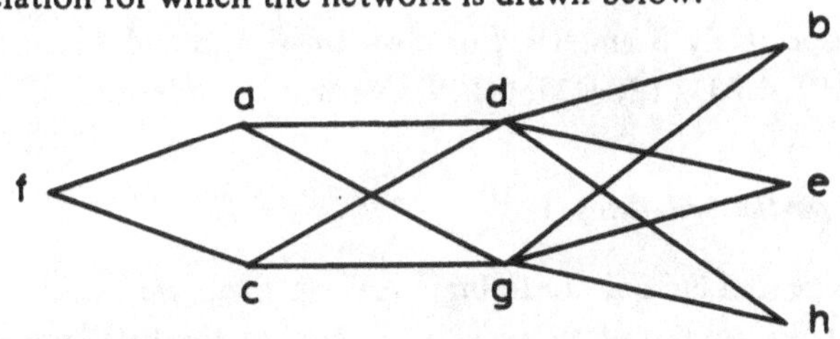

Fig. 23

To every ordering is thus associated a strict ordering. Conversely, to a strict ordering $<$ we associate an ordering relation $\leqslant$ by writing $x \leqslant y \Leftrightarrow x < y$, or $x = y$. There is thus a bijection between the set of orderings and the set of strict orderings on a set.

Note: a strict ordering has been defined as an antireflexive relation, that is, one that is antisymmetric and transitive. The reader should show that one of these properties is the consequence of the other two and can thus be omitted from the definition.

3.1.3 Duality

Let $(E, \leqslant)$ be an ordered set. We define on E a relation $\leqslant'$ by writing $x \leqslant' y \Leftrightarrow y \leqslant x \Leftrightarrow x \geqslant y$.
It is easy to show that $\leqslant'$ is an ordering relation, called the dual ordering relation of $\leqslant$.

For example, in $\mathcal{D}(E)$, the dual relation 'to be included in' ($\subset$) is 'includes' ($\supset$). In N, the dual relation of 'x divides y' is 'y is a multiple of x'. On the network representing an ordering we obtain the dual ordering by reversing the direction of the arcs (or by inverting the network).
Example:

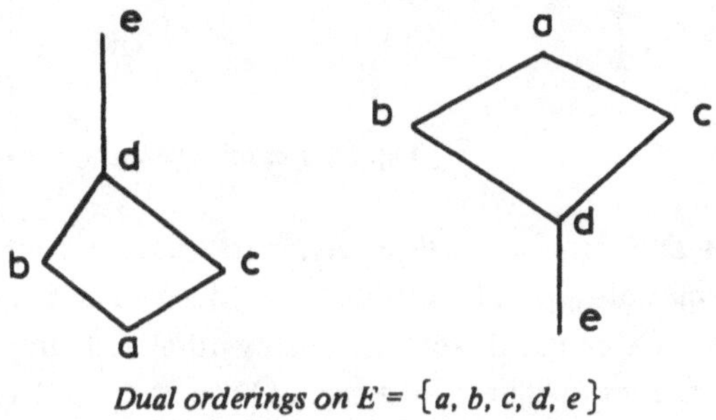

Dual orderings on $E = \{a, b, c, d, e\}$ Fig. 24

Example 8: Draw the network of all the ordering relations that can be defined on a set of two or three elements.

We find respectively 3 and 19. For a set of 4, 5, 6 and 7 elements, there are respectively 219; 4,231; 130,023; 6,129,859 distinct orderings.

3.2 Total and partial orderings

Orderings are particular quasiorderings. Among them we can distinguish quasi-orderings that are total and those that are not. In the first case we obtain the total orderings that satisfy the additional property:

$$\forall x \in E, \quad \forall y \in E, \quad x \nleq y \Leftrightarrow y \leqq x$$

In the second case we obtain partial orderings for which we have the property:

$$\exists\, (x, y) \in E^2 \quad \text{with} \quad x \nleq y \quad \text{and} \quad y \nleq x$$

We then say that x and y are two incomparable elements of the ordered set $(E, \leqslant)$. The relation of incomparability is sometimes written $x \| y$. But beware! By definition, the notation $x \leqslant y$ is equivalent to the notation $y \geqslant x$; but the notation $x \nleqslant y$ is not equivalent to the notation $y \leqslant x$ except for the total orders. A set on which a total ordering relation is defined is said to be totally ordered. If the relation is partial, the set is said to be partially ordered. In the latter case we sometimes call it an 'ordered set' without further distinction. Instead of saying that a set is totally ordered, we also refer to it as a chain or scale, particularly in the social sciences. For example, the set of whole numbers is partially ordered by the relation of divisibility, as the set $\mathcal{P}(X)$ of the subsets of a set X is also partially ordered by the relation of inclusion.

The network of a totally ordered set can be represented by aligned points as follows:

Fig. 25. *Representation of the total order $c < b < a$*

We see from this that for a set of three elements there are six orderings possible, obtained by permutating in all possible ways the elements in the network. In general, if E has n elements, there are as many total orderings on E as there are permutations on the set E, that is: $n! = n(n-1)(n-2)\ldots 3.2.1$.

There exist several characterizations of finite total orderings. Thus we will show that:

a strict total ordering ⇔ a total acyclic relation.

A relation R is total if for every pair x, y of elements of E, $x\ R\ y$ or $y\ R\ x$. A cycle for the relation R is a sequence of pairs:

$$(a_1, a_2), \ldots, (a_i, a_{i+1}), \ldots, (a_{n-1}, a_n) \text{ with}$$

(1) $a_1\ R\ a_2, a_2\ R\ a_3, \ldots, a_i\ R\ a_{i+1}, \ldots, a_{n-1}\ R\ a_n$, and
(2) $a_1 = a_n$.

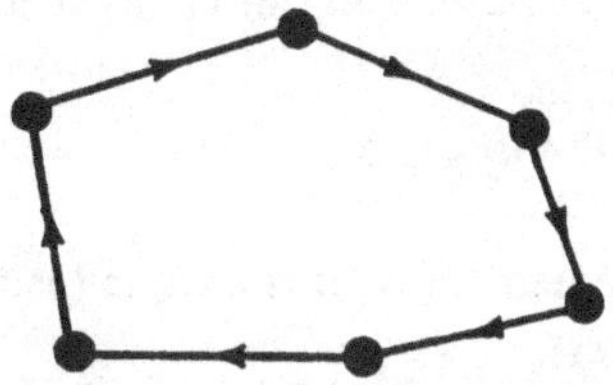

Fig. 26

If the cycle has n elements, it is said to be of order n. A strict ordering is an acyclic relation. Indeed, it follows from (1) by transivity that: $a_1\ R\ a_{n-1}$; whereas $a_{n-1}\ R\ a_n$ and $a_1 = a_n$ imply that a_1 and a_{n-1} are connected in the two directions, which contradicts the antisymmetry of the ordering relation.

Similarly, an acyclic relation is, in particular, antisymmetric (not a cycle of order 2) and antireflexive (not a cycle of order 1). Let us now show that if an acyclic relation is total as well, then it is also transitive.

Let x, y and z with $x\ R\ y$ and $y\ R\ z$. Since R is total, we also have $z\ R\ x$ or $x\ R\ z$. If $z\ R\ x$, then we would have a cycle of order three, which is impossible; thus, we have $x\ R\ z$ and the relation is transitive.

Exercise 9: Show that a strict (finite) total ordering is characterized as a total relation without cycle of order $\leqslant 3$.

In the case of a finite set E, all of the total orderings which we can define on E have the same structure (they are 'isomorphic'). Such is not the case for an infinite set. Thus, we are obliged, to distinguish between two types of orderings (see chapter VI, 'Measure scales', section 1).

Exercise 10: From the results of exercise 8 and the last example, deduce the number of partial orderings on a set of *2, 3, 4, 5, 6* and 7 elements.

This exercise shows that even for a set having few elements, there exist many partial orderings. We shall distinguish between the classes of partial orderings, and to do so we can use the definitions and results of the paragraphs to follow.

3.3 Particular elements of an ordered set

Let $(E, \leqslant)$ be an ordered set, and let A be a subset of E.

Upper bound: m, an element of E, is an upper bound of A if $\forall x \in A, x \leqslant m$. If A has at least one upper bound, A is said to be bounded above.

The greatest element (or maximum): m, an element of E is the greatest element of A if:

1. m is an upper bound of A,
2. $m \in A$.

If there exists a greatest element of A, it is unique (as follows from the property of the antisymmetry of order).

The maximal element: a, an element of E, is a maximal element of A if:

1) $\forall x \in A \qquad a \nless x$,
2) $a \in A$.

A subset A can contain zero, one, or many maximal elements; but if this subset contains a maximum then it is the only maximal element.

Universal upper bound: u, an element of E, is a universal upper bound if it is the greatest element of E. For example, in$(\mathcal{P}(E), \subset)$, E is the universal upper bound. On the other hand, $(N, \leqslant)$ does not contain a universal upper bound.

Copoint (or coatome): Let E be an ordered set for which there exists a universal upper bound u. Then, $x \in E$ is a copoint if u covers x: $x \prec u$. What are the copoints of $(\mathcal{P}(E), \subset)$?

In a similar fashion we define the notions of lower bound, of least element (or minimum), of universal lower bound (or null element), and of point (or atom). For example, the element m is the lower bound of the subset A if m is less than, or equal to, any other element of A. A point (atom) is an element that covers the universal lower bound (if it exists). What are the atoms of $\mathcal{P}(E)$, ordered by inclusion; of N ordered by divisibility? A subset that is bounded above and bounded below is said to be *bounded*.

We will now define the *least upper bound* (or supremum or join) of a subset A. Let us consider $M_A = \{\text{upper bounds of } A\}$. If there exists in M_A a least element, this element is called the least upper bound of A. In other words, the least upper bound s of A is the least of the upper bounds of A and thus satisfies the two conditions:

1. for any x of A, $s \geqslant x$; and
2. [for any x of A, $m \geqslant x$] $\Rightarrow m \geqslant s$.

It is clear that if the least upper bound exists, it is unique, and moreover, if A contains a greatest element m, it is the least upper bound.

Similarly, we define the greatest lower bound i (or infimum, or meet) of a subset A as the greatest of the lower bounds of A: for all x of A, $i \geqslant x$, and [for all x of A, $m \leqslant x$] $\Rightarrow m \leqslant i$

In the case where the subset A contains but two elements, $A = \{x, y\}$, the least upper bound, if it exists, is often written $x \vee y$, and sometimes as $\sup(\{x, y\})$. The least upper bound (often abbreviated as l.u.b.) of any subset, if it exists, is often written $\bigvee A$ or sup A. The same applies to the greatest lower bound (g.l.b.), if it exists, which is often written symbolically as $a \wedge v$, or $\bigwedge A$, or inf A.

Example:

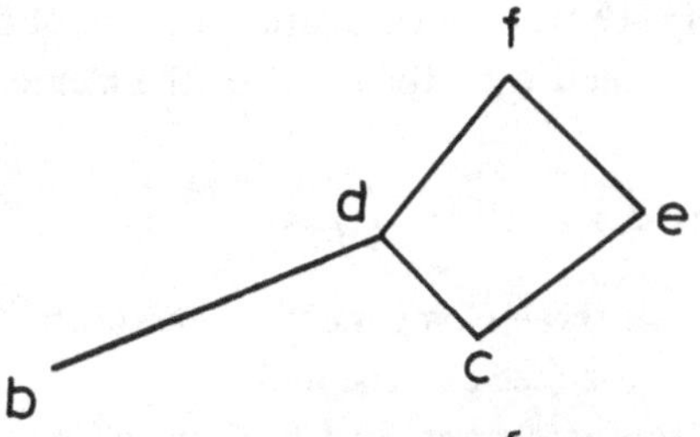

Fig. 27

$b \vee c = d$, $b \wedge c$ does not exist
$d \vee e = f$, $d \wedge e = c$
$b \vee f = f$, $b \wedge f = b$

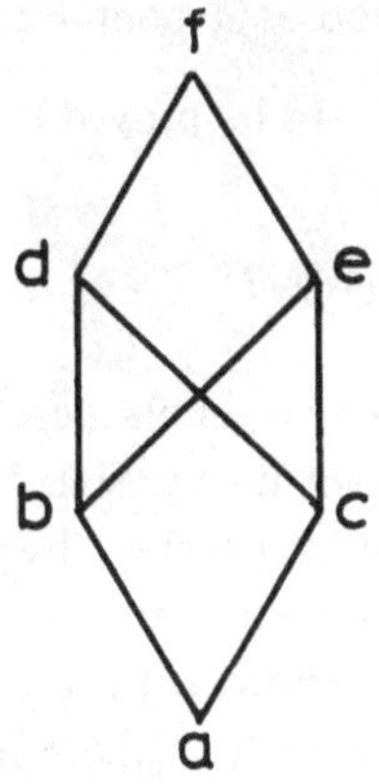

Fig. 28

$b \vee c$ does not exist
$b \wedge c = a$
$\inf(\{b, c, d, e\}) = a$

3.4 Semilattices, lattices and trees

A join-semilattice E *is an ordered set such that every pair of the elements of* E *has a least upper bound.*

Thus, in a join-semilattice, for any pair a and b of elements there exists an element s, written $(a \vee b)$ for which the following holds:

1) $(a \vee b) \geqq a$, $(a \vee b) \geqq b$,
2) $m \geqq a$, $m \geqq b \Rightarrow m \geqq (a \vee b)$.

For example, the ordered set below is a join-semilattice:

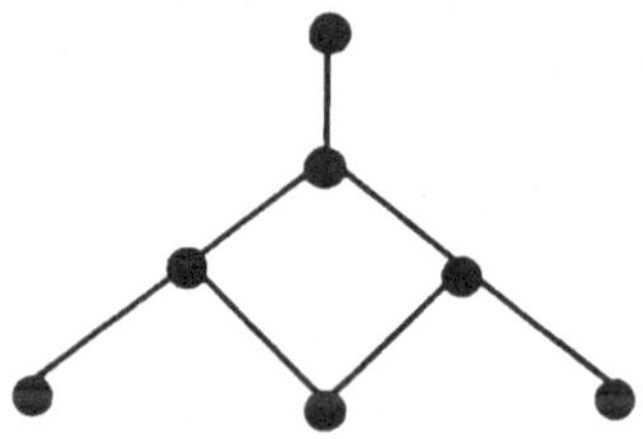

Fig. 29

Let $A = \{x, y, z\}$ be a subset of three elements of a join-semilattice E. Does A contain a least upper bound, sup A? We can calculate the l.u.b. of the elements two by two: $x \vee y$, $x \vee z$, $y \vee z$; and then the l.u.b. of the elements obtained with the third element:

$$(x \vee y) \vee z, \quad (x \vee z) \vee y, \quad (y \vee z) \vee x$$

The reader should show that these three elements equal the l.u.b. of the subset A; this l.u.b. is then written $x \vee y \vee z$ (see exercise 11).

More generally, any (finite) subset A contained in E has a l.u.b., written $x_1 \vee \ldots \vee x_i \vee \ldots \vee x_n$ or $\bigvee_{i=1}^{n} x_i$ (to be proved by induction). We note also that in E, we have the equivalence:

$$x \geqq y \Leftrightarrow x \vee y = x \quad \text{(Why?)}$$

Furthermore, we note that in a join-semilattice, the 'relational' property: one element is inferior to another, can be translated by an algebraic property: an equation. This makes it possible to develop, by algebraic means, the study of semilattices and explains their importance. It has been further shown that it is possible to define a semilattice uniquely by algebraic properties (see exercise 11). Moreover, the equational form which these relations assume make calculations easier (see for example, Boolean algebra, in chapter 3, 'Boolean algebras and Boolean rings').

Exercise 11: Given a join-semilattice, show that the operation which associates to two elements their l.u.b. is an associative, commutative and idempotent operation. Show conversely that in a set with an operation satisfying the three preceding properties, it is possible to define an ordering relation that generates a semilattice (use relation [1]).

Similarly, we define a meet-semilattice as an ordered set in which every pair of elements has a g.l.b. The dual of a join-semilattice is obviously a meet-semilattice, and conversely.

Finally, we define a lattice as a set being both join and meet-semilattices.

A lattice is an ordered set such that every pair x,y *of elements contains a l.u.b.* $x \vee y$ *and a g.l.b.* $x \wedge y$.

For example, $\mathcal{P}(X)$ ordered by inclusion is a lattice; the l.u.b. of A and B is $A \cup B$, and the g.l.b. is $A \cap B$. The set N of positive integers, ordered by divisibility, is a lattice where $x \vee y$ = least common multiple (x,y) and $x \wedge y$ = greatest common divisor (x,y). Every total ordering is obviously a lattice.

Example 12: From among the ordered sets of exercise 7 find the lattices and the semilattices.

Exercise 13: Show that every finite subset A of a lattice has a l.u.b., sup A and a g.l.b., inf A.

The lattices are particular semilattices; other semilattices of special interest are trees.

A tree is a join-semilattice for which every element, except the universal element, is covered by a single element.

Similarly, it is possible to define a tree as a meet-semilattice for which every element, except the null element, covers only one element. For example, the ordered set below is a tree:

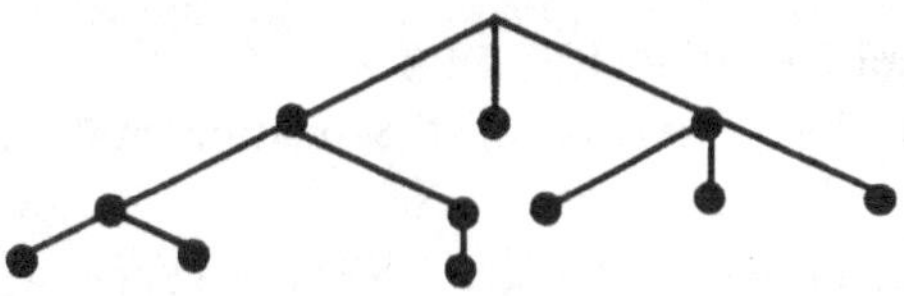

Fig. 30

In an ordered set we define a *chain* as a subset that is totally ordered. If $x \leqslant y$, we call an *interval*, $[x,y]$ the set of elements such that $x \leqslant z \leqslant y$. These definitions allow another charaterization of trees: a tree is an ordered set such that for every pair x, y, where $x \leqslant y$, the interval $[x,y]$ is a chain. In particular, a totally ordered set is a tree.

Exercise 14: Show that an ordered set that is both a tree and a lattice is totally ordered. We can thus represent the relations between semilattices, lattices, trees and total orderings as:

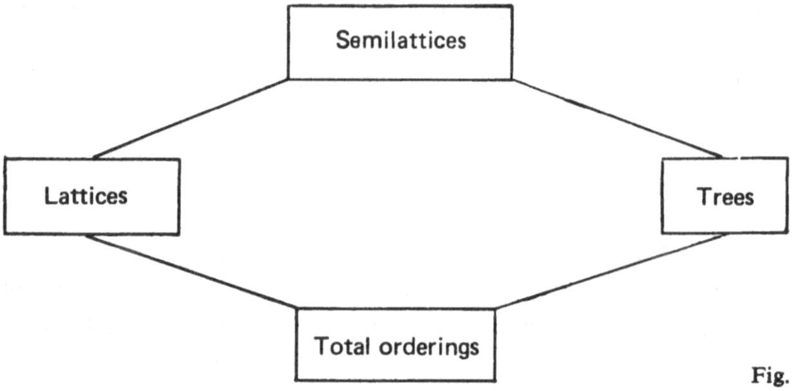

Fig. 31

A detailed study of trees is the subject of chapter 1, Trees[1] (see page 15)

4. Quasiorderings revisited

In the preceding sections we have studied two particular classes of quasi-orderings: the symmetric quasiorderings or equivalences, and the antisymmetric quasiorderings or orderings. We shall now show that any quasiorder reduces to the combination of an equivalence and an ordering.

Let (E, R) be a quasiordered set. Let us write:

$$x\,S\,y \Leftrightarrow [x\,R\,y \quad \text{and} \quad y\,R\,x]$$

Fig. 32

Lemma 1: S is an equivalence relation on *E*.
The reflexivity and the symmetry of the relation are evident at once, and

1. In this article a rooted tree or arborescence is what we call here a tree.

furthermore

$$x\,S\,y \text{ and } y\,S\,z \Leftrightarrow \begin{cases} x\,R\,y, \quad y\,R\,z \Rightarrow x\,R\,z \\ z\,R\,y, \quad y\,R\,x \Rightarrow z\,R\,x \end{cases} \Rightarrow x\,S\,z$$

Thus, S is transitive and is an equivalence relation on E. Let us now consider the quotient set E/S composed of the equivalence classes:

$$\{C_1 \ldots C_i \ldots C_p\}$$

In the set E/S we define a relation $\leqslant$ by writing:

$$[C_i \leqq C_j] \Leftrightarrow [\exists\, x \in C_i, \quad \exists\, y \in C_j \quad \text{with} \quad x\,R\,y]$$

Lemma 2. The relation $\leqslant$ is an ordering relation on E/S.
The reflexivity and transitivity of $\leqslant$ is easily prooved. Let us now show the following result:

$$[C_i \leqq C_j] \Rightarrow [\forall\, a \in C_i, \quad \forall\, b \in C_j, \quad a\,R\,b]$$
$$[C_i \leqq C_j] \Rightarrow [\exists\, x \in C_i, \quad \exists\, y \in C_j \quad \text{avec} \quad x\,R\,y];$$

Let $a \in C_i$ and $b \in C_j$; then we have $a\,R\,x$ and $y\,R\,b$, and finally $a\,R\,x$, $x\,R\,y$, $y\,R\,b$ $\Rightarrow a\,R\,b$ by the transitivity of R.

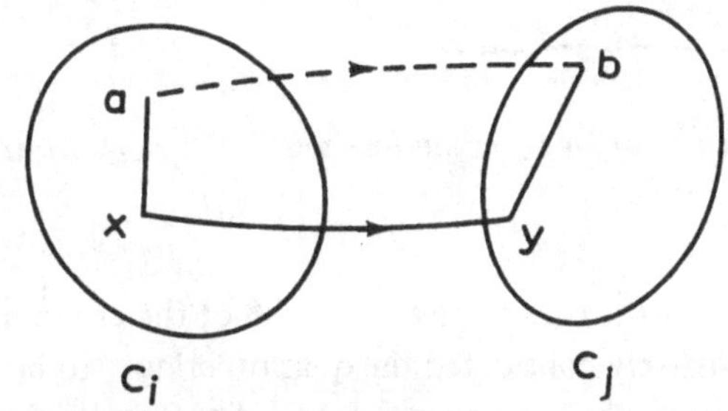

Fig. 33

The antisymmetry of $\leqslant$ remains to be shown. Let C_i and C_j be two distinct classes with $C_i \leqslant C_j$, $C_j \leqslant C_i$. Let $x \in C_i$ and $y \in C_j$. From the preceding result we can deduce $x\,R\,y$ and $y\,R\,x$, thus that $x\,S\,y$, which is impossible since we had supposed that x and y were in two distinct equivalence classes. Thus we have that E/S is ordered by $\leqslant$. The relation $\leqslant$ defined on the quotient set E/S is called the quotient ordering of the quasiordering R (by the equivalence S) and we write it R/S.

Lemmas 1 and 2 lead to the following result:

Theorem: Any quasiordering on a set is defined by the following:

1. an equivalence on the set; and
2. an ordering on the quotient set of classes defined by this equivalence.

Thus, we see that any quasiordered set can be partitioned into equivalence classes, these classes themselves being ordered. The notion of quasiordering thus formalizes the usual notion of classification; we begin by putting the set of equivalent objects together according to chosen criteria, and then we order these classes according to these criteria.

In the particular case where the quasiordering is symmetric, the ordering obtained is trivial (no class is comparable with any other). If the quasiordering is antisymmetric, the equivalence obtained is again trivial (each class is made up of a single element). If the quasiordering is total, the quotient ordering on the classes is total.

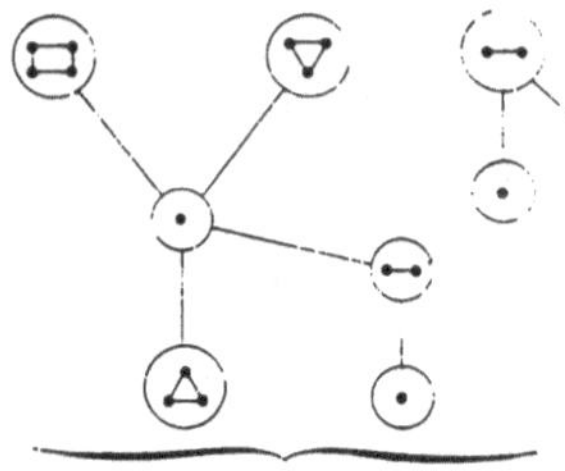

Fig. 34
Quasiordering of nine elements on a set of 21 elements

Fig. 35
Quotient ordering

Example 9: If we return to examples 2,5 and 8 of the group of students, we see that we have successively considered the quasiordering: "to be of the same age", and the strict ordering, "to be younger than". On the set of the age classes, the quasiordering induced a quotient ordering for which we give the following network:

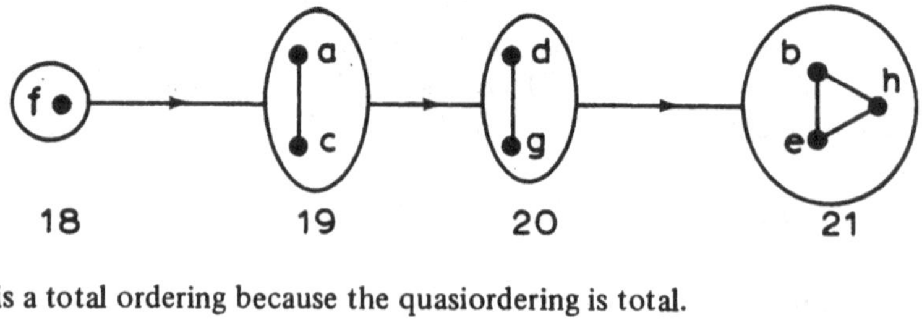

Fig. 36

This is a total ordering because the quasiordering is total.

Exercise 10: From the quasiordering of Example 3 we obtain the following quotient ordering: it is a lattice.

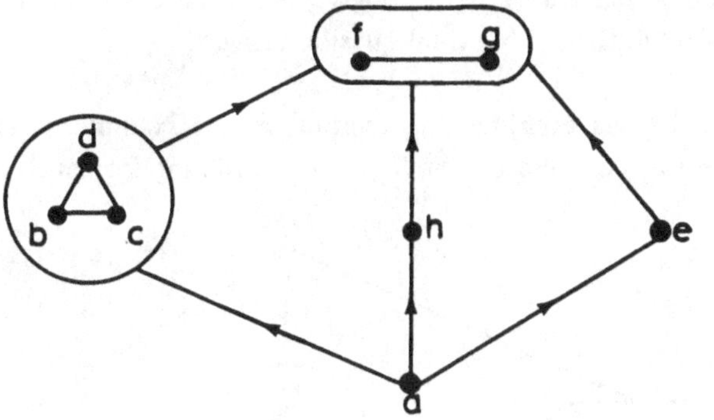

Fig. 37

5. Some applications

5.1 The transitive closure of a relation

In general, any relation R is not transitive. There exists, therefore triplets x, y, z with $x\ R\ y$, $y\ R\ z$ and $x \not R\ z$. Can we add pairs to the relation R to make it transitive? To add the pairs $x\ R\ z$ for all of the preceding triplets does not suffice (find a counter example); but we also wish to add as few pairs as possible. The solution exists and it is unique. It consists in constructing a relation R, the transitive closure of the relation R.

R is defined on the basis of the paths of R. A path for the relation R is a sequence of pairs $(x_1, x_2), \ldots, (x_i, x_{i+1}), \ldots, (x_{n-1}, x_n)$ with

$$x_1\ R\ x_2,\ \ldots,\ x_i\ R\ x_{i+1},\ \ldots,\ x_{n-1}\ R\ x_n$$

Such a path has the origin x_1 and the extremity x_n. Thus we can write:

$x\ R\ y \Leftrightarrow$ there exists a path of origin x and extremity y.

R is called the transitive closure or the relation R.

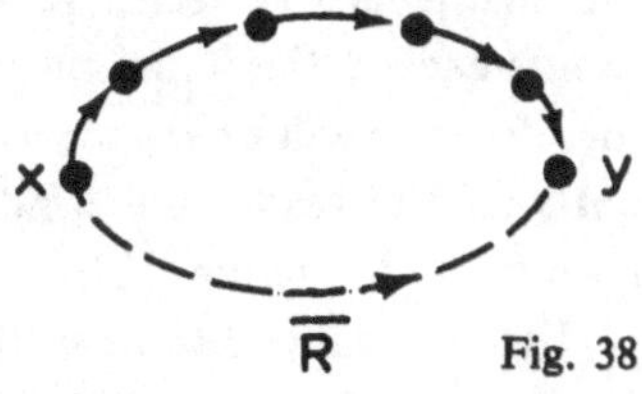

Fig. 38

Obviously we have $R \subset \overline{R}$ ($x\ R\ y \Rightarrow x\ \overline{R}\ y$). Moreover, $\overline{R}$ is transitive (why?). We will now show that $\overline{R}$ is the smallest transitive relation containing R (why?). In particular, if R is transitive then $R = \overline{R}$, and conversely.

If R is a reflexive relation, $\overline{R}$ is a quasiordering which we can thus decompose into an equivalence and a quotient ordering. The two elements are in the same equivalence class if, and only if, they belong to the same cycle (the equivalence classes are sometimes called the strongly connected components of R). If, in addition, R is total, then $\bar{R}$ is a total quasiordering.

Example 11: Let us consider the relation R of Example 1 assumed to be reflexive. The transitive closure of R is a quasiordering for which the network is given below:

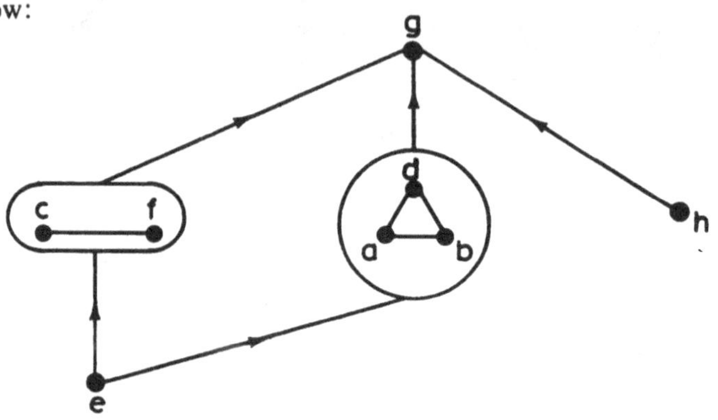

Fig. 39

To construct the transitive closure of a relation permits us, in general, to analyze it better as we saw in the example of the preceding sociogram. Thus, when presented with any relation, it is advisable to look for the transitive closure.

5.2 Other examples

The following examples present some situations where the notions of the preceding section will be used to construct mathematical models. We can thus use mathematical techniques and prove certain results. These techniques or results exceed the limits placed on this introductory material. For each subject, therefore we will limit ourselves to bibliographic references. It should be pointed out that the reader will find a general review of the mathematical methods in Barbut and Monjardet [2].

The first subject to be outlined here is that of the collective decision, or more precisely, voting procedures. We have n voters who must class p candidates. We presume that each voter expresses his individual preferences by a total ordering on the set of candidates. Thus, we obtain n total orderings, and the problem is to define a procedure for 'clustering' these total ordering in such a way as to define the collective preference of the voters.

A problem that is formally identical to the previous one is that of the 'choice with multiple criteria'. Here an individual must choose between several decisions for which the possible consequences are evaluated according to different criteria. For each criterion, we can totally order the consequences; we have thus several total orderings on the set of the consequences, and we must fuse them into a single total ordering. For these two problems consult Arrow [1]; Guilbaud [6]; and *La Décision* [10].

Ordering relations also apply to the analysis of questionaires. Thus, in hierarchic analysis we attempt to totally order the set of subjects having responded to a questionnaire, and the set of questions as well, in order to obtain the 'Guttman scales' see Matalon [8]. The study and generalization of the techniques used in the hierarchic analysis makes use of the properties of 'ordering lattices'. The same properties are useful for the development of certain problems raised in textual analysis (see Barbut and Monjardet [2], Degenne [4], and Frey [5]).

A strategy fundamental to all of science consists in classifying all of the data obtained. The archaeologist, for example, attempts to classify the hundreds of pot sherds he has found at a single site. A nuclear physicist attempts to classify the elementary particles, new examples of which are being constantly discovered. The problem common to all of these pursuits is to find a partition of the objects to be classified, or more generally, a sequence of partitions corresponding to finer and finer classifications. Thus, we are led to work with the 'partition lattice' of a set (see Barbut and Monjardet [2]). On the other hand, the desired classifications are established according to 'indicators of resemblance' between the objects in question. The indicators lead to the establishment of certain quasiorderings on the set of pairs of objects (see Lerman [7]).

In conclusion, it should be pointed out that many concrete problems can be formalized as follows: let E be a set, and R a binary relation on E (we sometimes say that the pair $G = (E, R)$ is a graph); find a partition of the set E for which the classes fulfill certain conditions with respect to the relation R. For example, the classes are the 'independant' subsets, that is: subsets for which any two elements of a class are not in relation. In addition, we often wish to find a partition satisfying the conditions imposed and for which the number of classes is maximum (or minimum). An example of such a problem is that of the colouring of a map using the minimum number of colours; the condition imposed is that two contiguous contries must not have the same colour. All of these problems are classic in the theory of graphs, but certain of them have not yet been solved (see Berge [3], and Matalon [8]).

Bibliography

1. Arrow, K. I., *Social choice and individual values.* New York, John Wiley, 2nd ed., 1963.
2. Barbut, M. and Monjardet, B., *Ordre et classification.* Paris, Hachette, 1970.
3. Berge, C., *Graphes et ses applications.* Paris, Dunod, 1970.
4. Degenne, A., *Techniques ordinales en analyse de données: statistique.* Paris, Hachette, 1972.
5. Frey, L., *Techniques ordinales en analyse de données: algèbre et combinatoire.* Paris, Hachette, 1972.
6. Guilbaud, G. Th., *Eléments de la théorie des jeux.* Monographies de recherche opérationelle, Paris, Dunod, 1968.
7. Lerman, J.-C., *Aspects théoriques de la classification automatique.* Paris, Gauthier-Villars, 1970.
8. Matalon, B., *L'Analyse hiérarchique.* Paris – La Haye, Mouton/Gauthier-Villars (Mathématiques et science de l'homme, vol. 1), 1965.
9. Ore, O., 'Theory of graphs', *Amer. Math. Soc. Colloq. Publ.* Providence. The American Mathematical Society, 1962, Vol. 38.
10. *La Décision: agrégation et dynamique des ordres de préférence,* Colloque international du C.N.R.S. Paris, C.N.R.S., 1969.

3. boolean algebras, boolean rings

by G. Th. Guilbaud

1. The present status of Boolean algebras

1.1 The Boolean algebras (it is better to use the plural) are in style. So much the better, but this is not without danger; the most mediocre popularizations have invaded the market, and since they sell, anything is printed. One of the reasons for this vogue is perfectly clear: the electronic techniques appled to networks, to computers, and more generally the organization techniques (operational research) need to augment the 'classical' mathematical tools with diverse algebraic structures that were not previously taught in 'class', and among these are the Boolean algebras.

Historical and social phenomena of this sort should be attentively examined. Mathematical instruction is sometimes very conservative, and the professional can, in retrospect, criticize his professors for having taught him mathematics for which he has little or no use, and for having neglected those disciplines that he needs today. But, paradoxically, it is not by persuing present trends that mathematical teaching will be most efficient: it is by presenting good mathematics, with neither blocks nor prejudices. This is one of the lessons of Boole.

More than a century ago George Boole in *The mathematical analysis of logic* (1847) said: 'They who are acquainted with the present state of the theory of Symbolical Algebra, are aware, that the validity of the processes of analysis does not depend upon the interpretation of the symbols which are employed, but solely upon the laws of their combination. Every system of interpretation . . . is equally admissible, and it is thus that the same process may, under one scheme of interpretation, present the solution of a question on the properties of numbers, under another, that of a geometrical problem, and under a third, that of a problem of dynamics or optics'. Boole also denounced another prejudice: that mathematics is but the science of magnitude, the science of the measureable! He pleaded for liberty in creative mathematics (cf. Nicolas Bourbaki, Eléments d'histoire des mathématiques [Paris, Hermann, 1969, 2nd. ed.], pp. 18 and 32.)

1.2 To apply these algebraic methods to the analysis of reasoning was not, even a century ago, an entirely new idea. Without going back to the middle ages and the combinatorics of Ramon Lull, one need only recall the works of Leibniz. A very simple example, will illustrate our trend of thought. In the *Generales Inquisitiones de Analysi Notionum et Veritatum,* which can be dated to 1686, we find the following effort:

> Let us represent the traditional forms of syllogism by a calculation. If I say that all men are mortal, I affirm that 'man is mortal' is redundant, and that the designation 'Man' suffices, because there are no other men but mortal ones.
>
> Let us try to write it
>
> $$HM = M \qquad (1)$$
>
> (where H means man (*homo*) and M means mortal).
>
> The same applies to a category of man, the Greeks if you wish; they are all men. Let us write it as
>
> $$GH = G \qquad (2)$$
>
> (The class G is that of 'Greeks').
>
> From these two equations, and by applying certain algebraic rules, I am able to 'deduce';
>
> – by associativity: $(GH)M = G(HM)$;
> – by substituting (1) and (2) in the preceding equation we have: $GM = GH$.
> – finally, by the transitivity of the equality, since $GH = G$ and $GM = GH$ we have:
>
> $$GM = G$$
>
> which can be interpreted in words as: 'all Greeks are mortal'.

Thus, here is an example of reasoning which can be represented by a manipulation of the algebraic type obeying certain precise rules.

1.3 As we have seen, this is not quite the algebra of our school children. In the algebra presented above, we must never 'simplify'. But how many professors think to put their students on guard against using the 'automaticisms' acquired in one algebra that are not necessarily valid for another?

There are some old jokes on this subject, but they bear retelling here. If dx/dy is a derivative, don't simplify by cancelling d. In the quotient $\sin x/\cos x$, don't simplify by cancelling x, etc.

1.4 It is said that we are the age of our arteries. I would say rather: 'We are the age of our algebras'; and it is the plural that counts. To go from one algebra to another is an excellent mental gymnastic. We sometimes complain of the bias seen in people due to an abuse of mathematics. The remedy is to learn more, and to diversify to know how to calculate for a group, a monoid, a ring, a field, or in a linear space, etc. and not to mix up the various operations. 'There are many who listen to the sermon in the same way as they listen to the Vespers', said Pascal, who had good training.

Perhaps one of the principal uses of the Boolean algebras is that they force us to break away from old habits, to limber our calculation reflexes. For this the Boolean algebras are particularly appropriate because they have a great simplicity and an aesthetic purity that is rather exceptionel.

1.5 However, we must not lose sight of the fact that these algebras are useful today in diverse professions. The two extensive domains (which in any case overlap) are computer science and formal logic.

It began with logic – first with Leibniz – but the important developments took place in the middle of the last century with Boole, De Morgan, Schröder, and others. Very early, the algebra necessary for logicians had become an object of study for mathematicians. In 1905 Couturat wrote: 'The fundamental laws of the algebra of logic had been invented to express the principles of reasoning *the laws of thought*, but one can consider these calculations from the purely formal point of view which is that of mathematics, as an algebra based on a few arbitrarily chosen principles (. . .) we demonstrate them not as logic but as algebra.'

Couturat's book *Algèbre de la logique* has been translated into several languages, among them, Russian, in 1910. P. S. Ehrenfest, in presenting this translation called the reader's attention to its possible applications. He pointed out that logical complexity is found in the construction of automatic telephone centrals and that the functioning of such a mechanism must be analyzed first from the qualitative point of view (all or nothing) and finally that the analysis of networks justified the use of such an algebra. This appeal took some time to be heard. Apart from a few isolated attempts, it was only from 1938 that electronic engineers began to use the Boolean algebras. The impetus to do so was apparently furnished by C. Shannon (U.S.A.), V. I. Chestakoff (U.R.S.S.) and A. Nakasima (Japan).

1.6 However, the use of Boolean algebras should not be restricted to logic, on the one hand, and electronics, on the other. We should understand that Boolean calculations have their place in numerous chapters of mathematics. A good part of what is called the algebra of sets is of Boolean form. The theory of measure and the theory of probability cannot do without Boolean algebra. When it comes to the teaching of Boolean calculations, the number of illustrations are overwhelming. But to begin, it would be better to present abstract examples. To speak of logic or electrical circuits, or even of probabilities would be perilous. On the other hand, today everybody knows what is meant by union, intersection, and complement, and this is a good way to introduce Boolean constructions.

Thus, we shall begin by taking a set U which will later serve as the universal set (the universe of the discourse, as the old logicians called it) which means that we will remain in this set. We know then that for any subset P of U (including the empty set and the whole set U) what is understood by the complement $\complement P$ or P^c of the subset P. For subsets P and Q we know how to define $P \cup Q$ and $P \cap Q$. And finally, we know what an equality is. Thus, we can begin to write equations such as:

$$(P \cap Q^c) \cup (P^c \cap Q) = (P^c \cup Q^c) \cap (P \cup Q).$$

We can propose a systematic study ot such equations after having taken note of the fact that all of the inclusion relations between subsets of the same set can be written in the form of equations since:

$$P \cap Q = P$$

just as

$$P \cup Q = Q$$

signifies that P is a part of Q.

To systematize this study, and to treat the significance of sets separately is to perform Boolean algebra.

We would do well to begin with a finite universal set (simplex) because when we consider infinite sets, a new phenomenon appears: we can conceive a system of subsets of U within which the three operations of union, intersection and complement are possible without leaving the system, even though the system does not contain *all* of the subsets of U. Let us consider, for example, the real line (that is to say, the set, ordered by $<$, of all the real numbers): starting from open intervals to the right and closed to the left:

$$a \leq x < b$$

and form all of the (finite) unions and intersections. We obtain those objects which form a Boolean algebra.

1.7 Thus, we will start with a finite universe. We know that the organization by inclusion of the subsets of a finite set is usually called a simplex. It is thus a question, more or less, of translating the ordered structure of a simplex into the form of an equation. What we obtain can be called a Boolean algebra.

Recall that if U possesses n elements, there are 2^n subsets, among which is the empty set 0. We begin with small values of the whole numbers and set up tables for the three operations: complement, union and intersection. The case where $n = 1$ is very easy. Let us therefore write the results for $n = 2$ by the following:

$U = \{a, b\}$; $A = \{a\}$; $B = \{b\}$; R = nothing, or the empty set.

Union	U	A	B	R
U	U	U	U	U
A	U	A	U	A
B	U	U	B	B
R	U	A	B	R

Intersection	U	A	B	R
U	U	A	B	R
A	A	A	R	R
B	B	R	B	R
R	R	R	R	R

	U	A	B	R
Complement	R	B	A	U

It is left to the reader to construct the tables for $n = 3$.

For each value of the integer n we can thus construct an algebra containing three operations and 2^n elements. Two algebras having the same cardinal are isomorphic. On the other hand, all algebras contain others (subalgebras) that are smaller, Thus, the preceding tables are still valid if we represent them (or interpret them) as:

$U = \{a, b, c, d\}$; $A = \{a, b\}$; $B = \{c, d, e\}$; R = nothing.

Therefore, we can deduce, in a certain measure, that the algebraic properties are independant of the interpretation that we give them.

But the objects manipulated by such algebras can be interpreted in still another and totally different fashion. For example, we can represent A and B as propositions: to remain in the mathematical domain, A will represent the phrase written in abbreviated form: '$x \leqslant y$'; while B will be written '$y < x$'. The universe U will be '*A or B*', and its complement R will be '*A and B*'. We note further that the use of the traditional logical conjunctions, 'and' and 'or', and the negation provides us with a satisfactory interpretation.

We are now in a position to study the algebraic aspect by temporarily making an abstraction of all interpretation.

2. Description of a Boolean algebra

A Boolean algebra contains objects and operations. These objects comprise a set which is given an algebraic structure by three operations.

2.1 We are thus given a set: all the elements of this set will be matched paired and we shall usually say that the two elements thus associated are the 'complement' of each other. If one of these elements is written A, its complement will be written A^c, for example.[1]

We note that if $A^c = B$ (if the object designated by the symbol B is the complement of the one designated by A), then: $B^c = A$. We call this an involutive correspondance (or a duality).

2.2 We now define two operations, both being commutative and associative, and mutually distributive.

Here again, it is recommended that matched paired symbols be chosen. Rather than imitate Boole who used the two crosses (+ and x) borrowed from traditional arithmetic, we can use parts of usual signs, half rounds, for example: $\cup$ and $\cap$, or half crosses such as: $\wedge$ and $\vee$, or $\top$ and $\bot$. In this article we will use the symbols which seem to be the most convenient in terms of manuscript: $\wedge$ and $\vee$. These symbols can be read as inferior or superior (inf and sup) but it is still easier to use the interpretation from formal classical logic: the two *conjunctions* 'or' and 'and'.

1. Here the diversity of usage is very great; some authors write A, others A' or A^* or even $-A$ $\neg A$, $\sim A$, etc.

We can thus write the following axioms:

1) A and B being any two elements, there exists an element C such that $C = A \wedge B$, and an element $D = A \vee B$.

2) $A \wedge B = B \wedge A$
   $A \vee B = B \vee A$ (commutativity)

3) $(A \wedge B) \wedge C = A \wedge (B \wedge C)$
   $(A \vee B) \vee C = A \vee (B \vee C)$ (associativity)

4) $(A \wedge B) \vee C = (A \vee C) \wedge (B \vee C)$
   $(A \vee B) \wedge C = (A \wedge C) \vee (B \wedge C)$ (distributivity)

Naturally, it would be easy to extend the preceding properties to the case where the expressions contain more terms. Thus, $A \wedge B \wedge C \wedge D$ designated an element independent of the order in which the expression is written as well as the way in which the operations are grouped. Similarly, there would be no difficulty in developing such forms as: $(A \wedge B) \vee (C \wedge D \wedge E)$ by repeated applications of distributivity.

2.3 Boole's law (also called tautology or absorption)

1. Repetitions are of no importance; we can write $A \vee A = A \wedge A = A$ for any A (idempotence).
2. It follows that if we have $A \wedge B = C$, for example, we can also write $A \wedge C = C$ and that $B \wedge C = C$.
3. We can add that $A \vee C = A$ and that $B \vee C = B$, and naturally, analogous rules in the opposite sense: $(A \vee B) \wedge A = A$ and $(A \vee B) \wedge B = B$ for any A and B.
4. Here again are easy generalizations such as: $(A \vee B \vee C \vee D) \vee A = A \vee B \vee C \vee D$ and $(A \vee B \vee C \vee D) \wedge A = A$.

2.4 Morgan's law (or the duality law)

If we have any equality, $A \wedge B = H$, for example, we obtain another equality by taking the complements of all of the elements, and reversing the operation symbol:

$$A^c \vee B^c = H^c$$

We can also write this law in the form:

$$(A \wedge B)^c = A^c \vee B^c$$

$$(A \vee B)^c = A^c \wedge B^c$$

Naturally, this rule holds for more complicated expressions as well.

2.5 Distinguished elements

Among the elements of a set, two of them (forming a complementary pair) are distinguished. We can call them the two poles of Boolean algebra. Some people noted them as did Boole himself, by the numbers 0 and 1. But it can be useful to use special signs; following the suggestion of Peano, a number of authors today choose $\bigwedge$ and $\bigvee$. We shall then first write:

$$\bigwedge^c = \bigvee \quad \text{and} \quad \bigvee^c = \bigwedge$$

Each of these two poles is neutral for an operation (it is for this reason that Boole used 0 and 1 with + and X):

$$\bigwedge \vee A = A \quad \text{and} \quad \bigvee \wedge A = A$$

But each of these is absorbant for another operation:

$$\bigwedge \wedge A = \bigwedge \quad \text{and} \quad \bigvee \vee A = \bigvee$$

Finally, we have for any element A:

$$A \wedge A^c = \bigwedge \quad \text{and} \quad A \vee A^c = \bigvee$$

An advantage of these notations is that they underline the relationship between the two operations ($\wedge$ and $\vee$) and the two poles ($\bigwedge$ and $\bigvee$). To summarize the preceding description we can set up the following table:

(commut.)	$A \wedge B = B \wedge A$	$A \vee B = B \vee A$
(ass.)	$(A \wedge B) \wedge C = A \wedge (B \wedge C)$	$(A \vee B) \vee C = A \vee (B \vee C)$
(dis.)	$(A \wedge B) \vee C = (A \vee C) \wedge (B \vee C)$	$(A \vee B) \wedge C =$ $= (A \wedge C) \vee (B \wedge C)$
(taut.)	$A \wedge A = A$	$A \vee A = A$
(Boole)	$(A \wedge B) \vee A = A$	$(A \vee B) \wedge A = A$
(Morgan)	$(A \wedge B)^c = A^c \vee B^c$	$(A \vee B)^c = A^c \wedge B^c$
(poles)	$\bigwedge^c = \bigvee$	$\bigvee^c = \bigwedge$

(neu.)	$A \wedge \bigvee = A$	$A \vee \bigwedge = A$
(abs.)	$A \wedge \bigwedge = \bigwedge$	$A \vee \bigvee = \bigvee$
(compl.)	$A \wedge A^c = \bigwedge$	$A \vee A^c = \bigvee$

2.6 *Axiomatics*

The propositions we have just been given are not independent. One can attempt to choose from the preceding list the smallest possible number of properties that make it possible to deduce the others; that is, a system of axioms.

This can be done in many ways, and since Huntington (1904) some twenty systems have been proposed. Here, for example, is a system which has the advantage of not concealing the duality;

1. Two operations, that is to say: the existence and uniqueness of the element written $A \wedge B$ and of the element written as $A \vee B$, for any elements A and B.
2. Distributivities:

$$A \wedge (B \vee C) = (A \wedge B) \vee (A \wedge C)$$
$$A \vee (B \wedge C) = (A \vee B) \wedge (A \vee C)$$
$$(A \vee B) \wedge C = (A \wedge C) \vee (B \wedge C)$$
$$(A \wedge B) \vee C = (A \vee C) \wedge (B \vee C)$$

3. The existence of neutrals: for any A we have:

$$A \wedge \bigvee = \bigvee \wedge A = A$$
$$A \vee \bigwedge = \bigwedge \vee A = A$$

4. The existence of a complement: for any A there exists an element X that is a solution of the system:

$$\begin{cases} A \wedge X = \bigwedge \\ \text{and} \\ A \vee X = \bigvee \end{cases}$$

From these postulates we can demonstrate the following theorems (and it is an excellent exercise):

1. If B is the complement of A, and if C is the complement of B, then C is non other than A.
2. Thus, the complement of A is unique.

3. The relations of tautology and absorption hold.
4. The operations are commutative.
5. The operations are associative.

It is possible to construct, without too much difficulty, as many Boolean algebras as we wish. But if our understanding is to be on a more profound level, it is advisable to use methods stronger than the simple combinatorial manipulations. We should not content ourselves either with simply the relations between the elements of an algebra, but should rather consider the algebras themselves as objects, and study the relations between the algebras. Without going as far as that (which would in itself merit the label of modern algebra), let me indicate some important perspectives.

Let us consider a Boolean algebra. To make our notions more concrete, we will use the algebra of two or four elements. Let us take the latter case: $(\bigwedge, \bigvee, A, B)$. Let us now take any set E and study the mappings of E into our algebra – that is, to consider the different ways of 'sticking' Boolean tags on all of the elements of E. Thus, if we choose E to be the set $\{1, 2, 3, 4, 5\}$ there will be four to the fifth power, or 1,024, possible mappings. Each of them could be designated by the list of Boolean values: thus,

$$f = (\bigwedge, A, A, \bigvee, B)$$

stands for the function (or mapping) which takes the value $\wedge$ in 1, the value $\vee$ in 4, the value A in 2 and 3, and the value B in 5. If we have another function

$$g = (A, \bigwedge, \bigvee, B, B)$$

we can easily define the Boolean operations $f \wedge g$ and $f \vee g$ by calculating each component separately (as with vectors). Thus:

$$f \wedge g = (\bigwedge, \bigwedge, A, B, B)$$
$$f \vee g = (A, A, \bigvee, \bigvee, B)$$

and similarly:

$$f^c = (\bigvee, B, B, \bigwedge, A)$$

Thus, it is easy to demonstrate that the mappings f, g, etc. constitute the elements of a new Boolean algebra. Therefore, we have a method for constructing Boolean algebras, in particular an infinite one. (Take, for example, E to be the set of natural numbers.)

If we start with a minimum algebra $(\bigwedge, \bigvee)$, a mapping of E into this algebra will be a partition of E into two parts, and we could find the algebra of the subsets of the set E without difficulty (the union and intersection being represented by $\vee$ and $\wedge$).

3. Boolean rings

3.1 Recalling that a set becomes an abelian group (or just *abelian* for the initiated) as soon as we are provided with an associative and commutative operation (most often noted by the '+' of addition) such that in every equation, $a = b + c$, knowing the value of two elements always determines the third. By considering the equation, $a + z = a$, we can show that its unique solution z, is a *neutral* element of the system, that is: $a + z = a$ implies $b + z = b$. We customarily call this neutral element 'zero'. Finally, the consideration of the equation $a + y =$ zero leads to the definition of the symmetry operation (often denoted, as in elementary arithmetic, by the 'minus' sign).

3.2 To obtain a *ring*, we must first have an *abelian*, to which we will assign a second operation, most often called multiplication, which is associative (but not always commutative), and is, in any case, distributive, that is:

$$a\cdot(b+c) = (a\cdot b)+(a\cdot c)$$

as is its left analogue. If we apply the distributive rule: $a(b$ + zero) we can prove without difficulty that for any a, we can write:

$$a\cdot\text{zero} = \text{zero}$$

We say that zero (*neutral* for addition) is *absorptive* for multiplication.

3.3 For a ring to be *Boolean*, we must have a commutative and tautological multiplication. This latter constraint (we say that multiplication must be idempotent) signifies that, for any element x, we have: $x \cdot x = x$. Along with most other authors, we shall require the existence of what is currently called a *unity*. That is, an element which will be neutral with respect to multiplication.

3.4 In any Boolean ring, in the meaning that has just been defined, there are thus two distinguishable elements which we can call 0 and 1, and which satisfy, for any p, the conditions:

$$0+p = p+0 = p$$

$$0 \cdot p = p \cdot 0 = 0$$

$$1 \cdot p = p \cdot 1 = p$$

from whence, in particular, come the fragments of the addition and multiplication tables for any Boolean ring:

+	0	1
0	0	1
1	1	

·	0	1
0	0	0
1	0	1

and these can be completed by the following remark: by the axiom of tautology (or idempotence) we have:

$$(1+p)\cdot(1+p) = 1+p$$

which by distributivity gives us:

$$1\cdot 1+p\cdot 1+1\cdot p+p\cdot p = 1+p$$

that is (1 is neutral for multiplication):

$$1+p+p+p\cdot p = 1+p$$

And finally, by the tautology and associativity:

$$(1+p)+(p+p) = (1+p)$$

which proves, finally that for any p:

$$p+p = 0$$

and in particular: $1 + 1 = 0$.
Thus, we have:

+	0	1
0	0	1
1	1	0

·	0	1
0	0	0
1	0	1

We have found that there exists a Boolean ring of two elements (it is the minimum possible); and we can quickly learn to calculate in this system since when using zero and one, the only thing that is new, different from the arithmetic of our childhood, is that we must say: 'one plus one equals zero'.

3.5 For certain applications this tiny Boolean ring suffices, from whence comes the illusion maintained by some non rigorous books and that Boolean calculation is non other than binary calculation.

But it is illuminating to examine the way in which Boolean rings of more than two elements are constructed, and to learn to calculate within them. The first thing to note is that we have no need of the 'minus' sign, which is necessary in most of the other rings.

Indeed, there is always for any p: $p + p = 0$, which comes to the same thing as saying that 'plus p' is equal to 'minus p', thus making 'to change sides' – the first gesture in abelian calculation – an easy step. Therefore, to solve $a + x = b$, we write:

$$a+x+x = b+x$$

that is:

$$a = b+x$$

Or better yet, if we are given: $a + b = c$, we can write $a + b + c =$ zero, and so forth.

3.6 Suppose that a Boolean ring contains, in addition to 0 and 1, another element. One might be tempted to choose 2. But the properties of this element are too different from the number two of the *ring* of ordinary integers for us to be unconcerned about the confusion that might result from its use (indeed, for some people, the use of 0 and 1 in Boolean calculation can be a source of confusion). Let us call the third element B (in Boole's honour). First we have:

+	0	1	B
0	0	1	B
1	1	0	?
B	B	?	0

·	0	1	B
0	0	0	0
1	0	1	B
B	0	B	B

We have yet to discover what is meant by: $1 + B$.

This is easy: it can be neither 0 nor 1, nor B not at least if B is different from 0 and 1. It can be easily proved (by recalling the abelian characteristic: in an equation such as: $x + y = t$, each element is uniquely determined by the two others).

Therefore, there is no Boolean ring having three elements, The existence of B makes the existence of a fourth element obligatory; this one can be designated as B', the companion of B, which satisfies: $1 + B = B'$, or further: $1 + B + B' =$ zero, and $B = 1 + B'$.

Therefore, we can set up the two tables:

+	0	1	B	B'
0	0	1	B	B'
1	1	0	B'	B
B	B	B'	0	1
B'	B'	B	1	0

$\cdot$	0	1	B	B'
0	0	0	0	0
1	0	1	B	B'
B	0	B	B	0
B'	0	B'	0	B'

Memorandum: We should note that for *addition*, 0 is neutral, as for the other members 1, B, B': the sum of two different elements equals the third, and the sum of two identical elements is 0.

We should note that for *multiplication*, 0 is absorbing, 1 is neutral, the square of an element is equal to itself, and the product of two 'companions', i.e. B and B', is equal to zero.

We have thus constructed a Boolean ring of four elements. (We will show in passing that all of the Boolean rings having four elements are isomorphic and only differ by the names given to the objects of which they are composed.)

3.7 Continuing to construct richer and richer Boolean rings in this manner is not impossible, but it is rather tedious. It can be proven that there exist rings of 8, 16, . . ., 2^n, . . . elements; that two Boolean rings having the same number of elements are isomorphic; that in any Boolean ring, the elements are paired; and that the companions are defined by:

$$1+B+B' = \text{zero},\ 1+B = B' \text{ and } 1+B' = B.$$

But if one wishes to have a more profound understanding of Boolean algebras it would be useful to use more powerful methods than the simple combinatoric calculations used in the preceding little examples. It would be necessary to turn to the general theory of rings (for which the relationships between the rings themselves are the objects of study, and not the relations between elements of the same ring). Without venturing too far in this direction (which in itself merits the name 'modern algebra') let us consider a few perspectives.

3.8 Let us consider a Boolean ring A – to be specific, we can take either a ring of two elements, or better yet, the ring 2^2, which we have just constructed – and any set E. Let us examine the mappings of E into A (that is, the different ways of assigning Boolean labels to the elements of E).

For example, if we choose E as the set of five elements $\{a, e, i, o, u\}$ and choose the ring of four elements $\{0, 1, B, B'\}$ as A, there are four to the power of five, that is, 1 024 possible mappings. Each one can be designated by the list of the values in A.
Thus:

$$f = (1, 0, 0, B, 0)$$

stands for the function (or mapping) which takes the value 1 in a; the value 0 in e, i and u; and the value B in o.

If we are given another function:

$$g = (0, 1, B', 1, B)$$

we can easily define the Boolean operations (by calculating them term by term):

$$\begin{aligned} f+g &= (1+0, 0+1, 0+B', B+1, 0+B) \\ &= (1, 1, B', B', B) \\ f \cdot g &= (0, 0, 0, B, 0) \end{aligned}$$

and prove that these definitions constitute a Boolean ring: the set of mappings $E \to A$.

It is then easy to construct as many Boolean rings as one desires, and in particular, infinite rings. Thus, we can take the set of natural numbers as the set E.

3.9 As soon as we have a Boolean ring, we have as many as we wish by considering the mappings of any set into the ring in question, and by supplementing the given set of these mappings with the necessary operation, as we have just seen for the preceding example.

If we start with the minimum ring, to map a set E into such a ring consists simply in assigning to each element of E the mark *zero* or the mark *one*, that is, to designate one subset of E marked 1, the rest being marked 0. There is thus a natural correspondence between the mappings of E into (0,1) and the subsets of E (this is the technique known as the characteristic function).

It is then simple indeed to relate the operations of a Boolean ring to the traditional operations of the algebra of subsets of a set.

It is easily proved that: $f \cdot g$ is the image of the intersection, and that $f + g$ is the operation that consists of selecting the elements which are contained in only one of the two subsets (symmetrical difference). It is thus that we have found *Boolean algebras*, but it is important not to confuse the two structures.

4. simplicial objects

by G. Th. Guilbaud

1. The first model: simple complex or simplex

1.1 Outline of an algorithmic description

1.1.1 Alphabet. *Words.* Elementary *operations* on Words.
1. Recopying without changes (orthographia).
2. Recopying and doubling a letter (dittographia).
3. Recopying and ommiting a letter (haplographia).

These operations are combined in every possible way starting with a *source* word.

1.1.2 Example
Diagrams are more readable when the resulting words are situated on different levels according to their lengths:

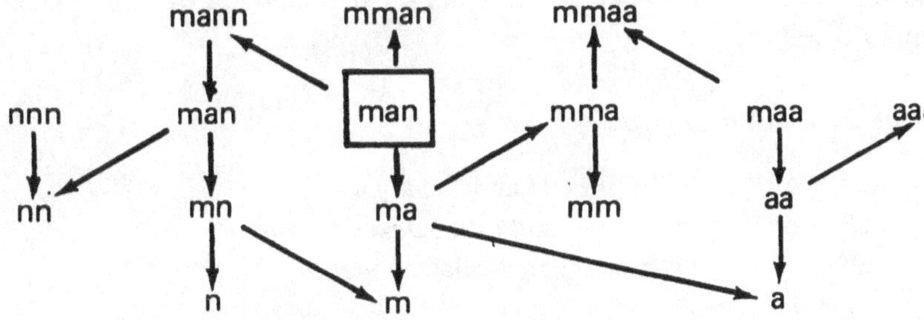

Fig. 1

Exercise: For the above diagram, draw all of the possible arrows.

1.1.3 The drawing above (Fig 1) illustrates only a fragment of what can be obtained by applying the elementary operations beginning with the source word, *'man'*. The totality of words that can be generated in this fashion is designated by the notation:

$$\Delta\,[man]\,,$$

and it is called a *simplex*.

The delta (Δ) evokes the triangle, the archetype of the simplex. If we limit ourselves to simplifying operations d (descending arrows) we obtain the classic diagram for the word *man*

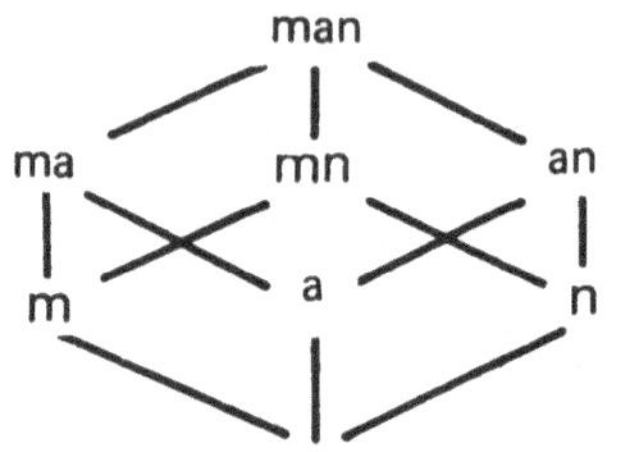

Fig. 2

1.2 Organization of the simplex

Premiminary remark: if all the letters of the source word are distinct, the structure depends only on the length of the word; from whence we have the abbreviated notation: $\Delta(3)$, $\Delta(4)$, etc. and $\Delta(w)$ for the cardinal w.

1.2.1 We start with a set of words which is partitioned into:

Δ^1 a set of words of one letter,
Δ^2 a set of words of two letters,

and so on to Δ^n. To these we add Δ^0, the set that contains but one element: the empty word.

1.2.2 The classes of mappings:

$\Delta^n \to \Delta^{n-1}$, written as d^n and called *faces*,
$\Delta^n \to \Delta^{n+1}$, written s^n and called *degenerations*,
$\Delta^n \to \Delta^n$, written o^n and called an *identity*

1.2.3 Diagram:

Arrows d:	$\Delta^0 \longleftarrow$	$\Delta^1 \leftleftarrows$	Δ^2 (three arrows $\longleftarrow$)	$\Delta^3 \ldots$
Arrows 0:		$\updownarrow$	$\updownarrow$	$\updownarrow$
Arrows s:		$\Delta^1 \longrightarrow$	$\Delta^2 \rightrightarrows$	$\Delta^3 \ldots$

Complete indexing

d_j^n = deletion of the jth letter in a word of n letters,
s_j^n = doubling of the jth letter in a word of n letters,
o^n = recopying of a word of n letters,
(these indices can ultimately be abridged to d_j, s_j, s^n, o^n, d, s, o).

1.2.4 Composition is associative. An operation d^n can be followed by o^{n-1}, d^{n-1}, or s^{n-1}; s^n can be followed by o^{n+1}, d^{n+1}, s^{n+1}; an o^n can be followed by o^n, d^n, s^n.

1.2.5 There are five types of fundamental relations that are true regardless of the length of the source word, and from which all of the others can be deduced.

1. The o are neutral, from whence the first types of equations such that: $od = d$ and $do = d$, etc.

2. The second type of form: $sd = o$ when a letter is doubled to be subsequently deleted.

   Examples: $s_1^n d_1^{n+1} = o^n$, $s_1 d_2 = o$, $s_i d_i = o$.

3. A composition of the form $d^n d^{n-1}$ can be written in two ways depending on the order in which two letters are deleted.

   Examples: $d_5 d_3 = d_3 d_4$ $d_j d_i = d_i d_{j-1}$ (if $i < j$).

4. The fourth type is analogous to the preceding: $ss = ss$.

5. The fifth type is $sd = ds$. When the repeated letter is different from the deleted letter it is possible to choose the order of the two operations because there is no interference.

   Example: $s_1 d_3 = d_4 s_1$.

   (See below 3.1.3 for the complete table of the relations.)

1.2.6 Note: When writing *ds, dd, sd,* etc. we adopt the convention that the operation written to the left of the equal sign is performed before the other.

1.2.7 Cannonical form of a composite operation

Any program of elisions and repetitions can be written with *d, o,* and *s* in several ways. But we can adopt the convention that we shall begin by first writing all of the *d*'s, and then *s*'s (or the contrary). We can also decide that the *d*'s will be placed according to the decreasing inferior indices (or increasing, according to one's preference), and similarly for the successions of the *s*'s.

Example:

$$\begin{aligned} d_3d_4d_7s_1s_5s_6s_8 &= d_9\, d_5\, d_3\, d_5\, s_4\, s_4\, s_1 \\ &= s_1\, s_7\, s_8\, s_{10}\, d_{13}\, d_6\, d_4 \\ &= s_7\, s_6\, s_6\, s_1\, d_4\, d_5\, d_{11} \end{aligned}$$

(an operation that transforms the word 'lucrative' into 'llurttiiv').

1.3 Traditional combinatorics

To study simplicial structures in detail according to the well-known dictum 'enumeration, numbering, organization' leads us to the classical rudiment: letter combinations, the arithmetic triangle of Pascal, and so forth. I will give here but a few exercises.

1.3.1 The sets Δ^n can be assigned a partial order generated by the order of the letters in the word source.

Example: We are given the word *PAUL;* the corresponding diagram of Δ^3 PAUL is given below with its 'natural' partial order.

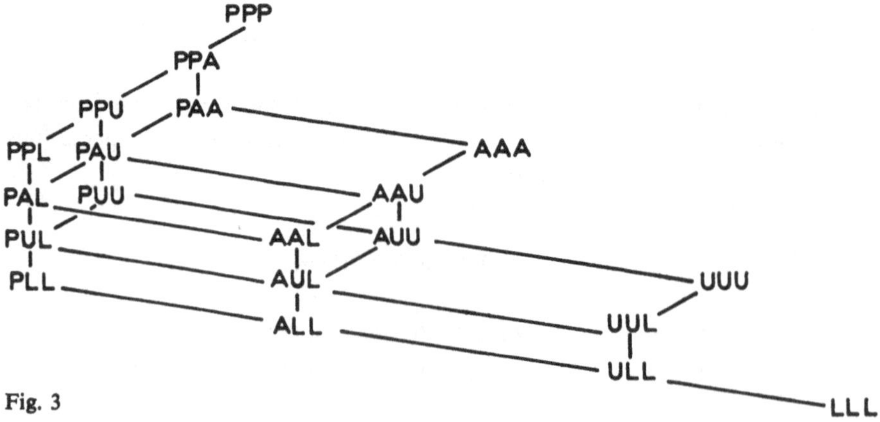

Fig. 3

Note: If the set of the four letters in the word 'PAUL' is designated by E, the diagram above is a subset of the cartesian product $E{\times}E{\times}E$ or E^3 (tetrahedron: a sixth of the cube).

The number of words: (cardinal) card $\Delta^n(m)$ is found in the Pascalian triangle (we used to say: it is the number of combinations, complete or with repetitions). In the drawing (Fig 3), in any case, we see the number in question is the triangular sum (Pascal called it pyramidal):

$$10 + 6 + 3 + 1 = 20,$$

and by another formula:

$$\text{card. } \Delta^3 (4) = (4.5.6)/(3.2.1) = 20.$$
$$\text{card. } \Delta^n(m) = \{m(m+1)(m+2)\ldots\}/\{n(n-1)(n-2)\ldots\}$$

(where there are as many whole consecutive factors in the numerator as in the denominator).

Another exercise is to draw Δ^4 (PAUL) which contains:

$$\frac{4.5.6.7}{1.2.3.4} = 35 \text{ elements.}$$

1.3.2 Let us now enumerate, name and organize the set of the mappings $\Delta^4 \to \Delta^3$: not only the four simplifications noted d_1^4, d_2^4, d_3^4 but also all of the mappings obtained by combining the d's and the s's in all possible ways. We can use the results of (1.2.7), by choosing one of the cannonical forms. The possible forms for a mapping $\Delta^4 \to \Delta^3$ are:

$$d^4, \qquad d^4\, d^3\, s^2, \qquad d^4\, d^3\, d^2\, s^1\, s^2;$$

and it only remains to choose the inferior indicies (decreasing for d, indreasing for s).

It is an easy exercise to show that there are 20 possibilities. But it is easier still to simply observe that the organization of the set of words $\Delta^3(4)$ can serve as well for the set of mappings $\Delta^4 \to \Delta^3$.

Therefore, let f be a mapping, f: $\Delta^4 \to \Delta^3$.
This mapping is distinguishable by the effect that it has on an element of Δ^4, provided that we take the precaution of choosing a word in which all of the letters are distinct.

That is, we will consider f (PAUL), a word of three letters chosen from Δ^3 (PAUL).

The correspondence between $\Delta^3(4)$ and $\Delta^4 \to \Delta^3$ is thus defined and obviously bijective.

1.3.3 Another useful similarity: a word can be identified with the mapping of a totally ordered set in an alphabet. Thus, we can have a word of m letters: a mapping of $(1 < 2 < 3 < \ldots < m)$ in an alphabet.

We are thus led to the study of the monotonic mappings (increasing, generally) of $(1 < 2 < \ldots < m)$ in $(1 < 2 < \ldots < n)$ to which correspond the mappings $\Delta^n \to \Delta^m$ (note the reversal of the arrow).

2. Some other models

2.1 Let us consider a monoid, and to avoid complications, the simplest subset of all: the monoid of the integers, the school child's addition tables and take only a small part of it:

$$3 + 4 \to 7.$$

As an illustration of associativity, a fundamental characteristic here, we explain that: $1 + 2 + 4$ can 'give' either $3 + 4$ or $1 + 6$; and that finally, the result is the same. We can illustrate this with a diagram:

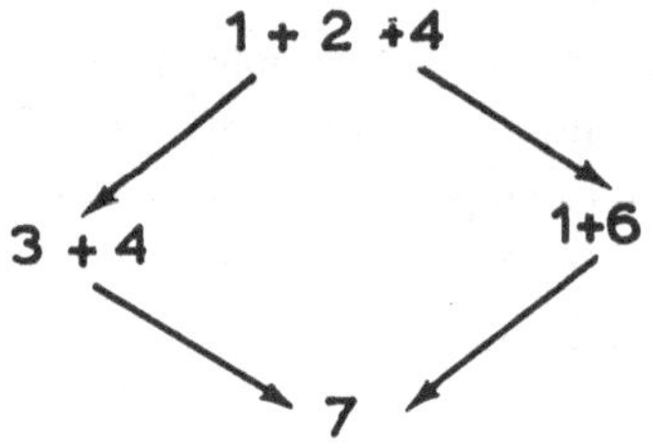

Fig. 4

This result can be generalized; below is the case of the sum of four terms:

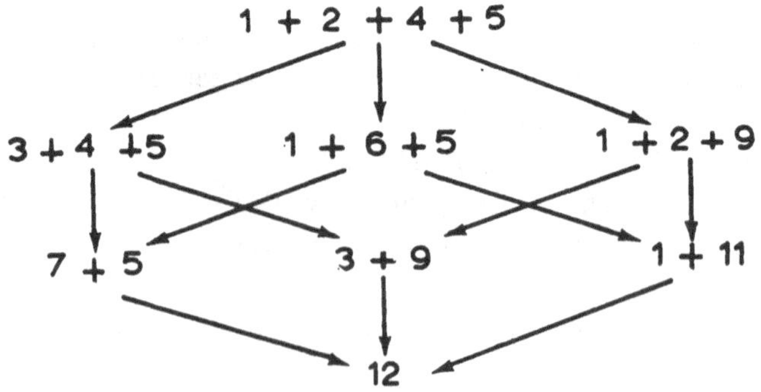

Fig. 5

Thus, we have found the well-known simplex diagram.

2.2 Let M be a monoid. The operation will be noted as the '+' for the whole numbers (but we do not assume that this operation is commutative, only associative). Let $x = (x_1, x_2, \ldots, x_i, \ldots, x_n)$ be a list of the elements of M. Since each x_i is a member of M, x_i is a member of the cartesian product M^n. Let us now introduce the mappings $d_i : M^n \to M^{n-1}$ which transform (x) into (y) by:

$$y_1 = x_1,\ y_2 = x_2,\ \ldots,\ y_i = x_i + x_{i+1},\ y_{i+1} = x_{i+2},\ \ldots$$

It is easy now to give the rule for the composition of two operations (d).

2.2.1 First case: Two additions of two pairs of neighbouring terms, but the two operations are distant enough so that they do not interfere.

This transformation can be obtained in two different ways because the order in which the two additions are effected can, obviously, be different.

Thus, let:

$$(\ldots, x_i, \ldots, x_j, \ldots) \quad \text{become} \quad (\ldots, x_i + x_{i+1}, \ldots, x_j + x_{j+1}, \ldots),$$

which can be obtained either by:

first d_i, and then d_{j-1}

or

first d_j and then d_i.

2.2.2 Second case:

$$(\ldots, x_i, x_{i+1}, x_{i+2}, \ldots) \qquad \text{becomes} \qquad (\ldots, x_i + x_{i+1} + x_{i+2}, \ldots),$$

which can be decomposed into d_i followed by d_i, or into d_{i+1} followed by d_i.

Here again we find the rules already seen in paragraph (1.2.5.3) above, and we should note the relation between these rules and the associative law.

2.2.3 When the monoid in which we are working is a group, we may use a well-known procedure of the statisticians and the probabilitalists, which consists of replacing the given list $x = (x_1, x_2, \ldots, x_n)$ by that of a 'cumulative' list $X = (X_1, X_2, \ldots, X_n)$ defined by:

$$X_1 = x_1, \qquad X_2 = x_1 + x_2, \qquad X_3 = x_1 + x_2 + x_3, \text{ etc.}$$

and also:

$$x_1 = X_1, \qquad x_2 = X_2 + \overline{X}_1, \qquad x_3 = X_3 + \overline{X}_2, \text{ etc.}$$

($\overline{X}$ is the opposite of $X : X + \overline{X} = 0$).

We thus have bijections between the x's and the X's.
The operation d_i transforms x into y:

$$y_h = x_h \text{ si } h < i,\ y_i = x_i + x_{i+1},\ y_j = x_{j+1} \text{ si } i < j,$$

resulting in a transformation of X into Y:

$$Y_h = X_h,\quad Y_i = X_{i+1},\quad Y_j = X_{j+1}.$$

Which means, finally that:

$$(\ldots, X_{i-1}, X_i, X_{i+1} \ldots) \to (\ldots, X_{i-1}, X_{i+1}, \ldots)$$

The operation d_i in the X space is non other than the deletion of a coordinante (projection) as in the first model. This leads to the idea of defining the operation s_i first on the cumulatives X: $(X_1, \ldots, X_n) \to (\ldots, X_{i-1}, X_i, X_i, X_{i+1}, \ldots)$ by repetition of X_i.

Moreover, we can deduce the definition of s_i for the x's:

$$s_i : (x_1, \ldots, x_n) \to (x_1, \ldots, x_i, 0, x_{i+1}, \ldots, x_n)$$

by writing 0 for the neutral element of the group.

2.3 Let us return to the monoid. The operations d: $M^n \to M^{n-1}$ having been defined, it only remains to define the s's. We wish first (see 1.2.5.2 above) that $s_i d_i$ will be identical to the transformation $s_{i-1} d_i$.

The results of (2.2.3) suggest a solution: the operation s_i consists in interdigitating a neutral element (recall that every monoid must have a neutral). We can then show all of the relations in (1.2.5) without difficulty.

Example: The integers:

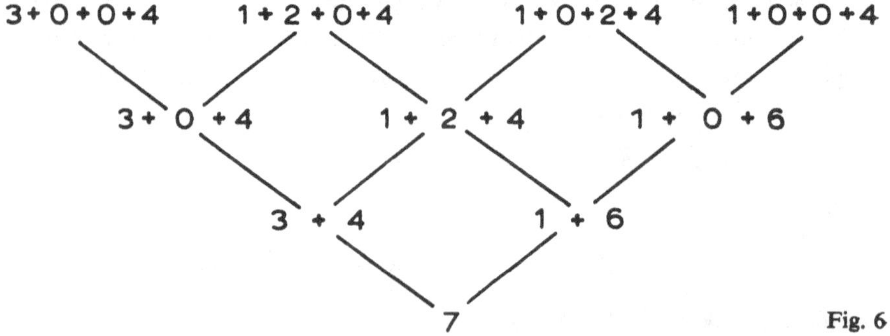

Fig. 6

have the same form as the simplex:

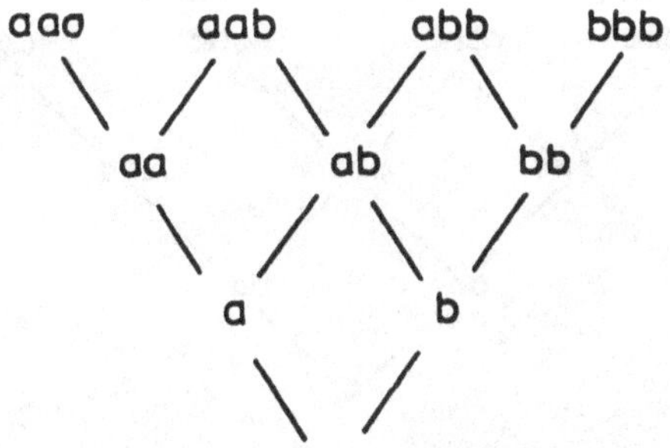

Fig. 7

2.4 But it is not even necessary to insist on a monoid because all that was needed for the preceding construction was: 1. associativity; and 2. the existence of neutrals.

Thus, we need only begin with a *category*, that is, an associative algebraic structure, but whose composition is not everywhere defined.

For the moment, let us be satisfied with the usual intuitive images and their *graphic* representations.

Let us consider three arrows, or three paths, end to end:

We can combine a and b, as well as b and c, and write:

$$ab = p, = bc = q$$

Similarly,

$$pc = aq = t$$

As for the interdigitated neutrals, they are the 'nul' paths, here depicted as intermediate 'stations':

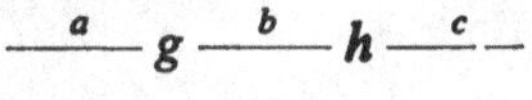

and we can write:

$$ag = a \quad gb = b \quad bh = b \quad hc = c$$
$$gq = q \quad ph = p$$

Finally, all the algebraic tools can be described by the operations (d) and (s)

performed on the 'words' and which can be summarized on the same sort of diagram:

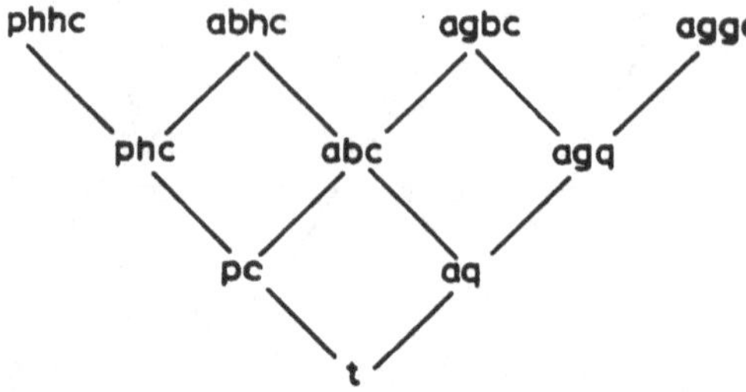

Fig. 8

2.4.1 It may seem surprising that only intermediate neutrals have been introduced, and that the extremities have been neglected:

$$f \xrightarrow{a} g \xrightarrow{b} h \xrightarrow{c} i$$

It is true that nothing prevents us from defining, for the word '*abc*', not two but four operations *s* which are respectively:

$$abc \longrightarrow fabc,\ agbc,\ abhc,\ abci$$

But if we wish to scrupulously conserve the simplicial form, it will be necessary in addition to define four simplifying operations of the class *d*. Naturally, all of the formal properties must be conserved as well.

This is not difficult. We will show this as an exercise for which the following is a good solution:

$$d_1(abc) = bc,\ d_2(abc) = pc,\ d_3(abc) = aq,\ d_4(abc) = ab$$

Or, expressed in words, the operations d_i consist in comparing two neighbouring letters of the given word, except at the two extremities where the operations are the deletion of the first and the last letter respectively.

Below we have the resulting variation on our old model of the addition of two integers. The three operations d^3 are:

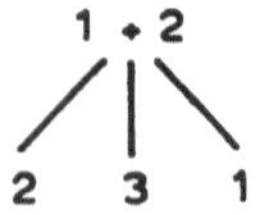

Fig. 9

And for the three s^3:

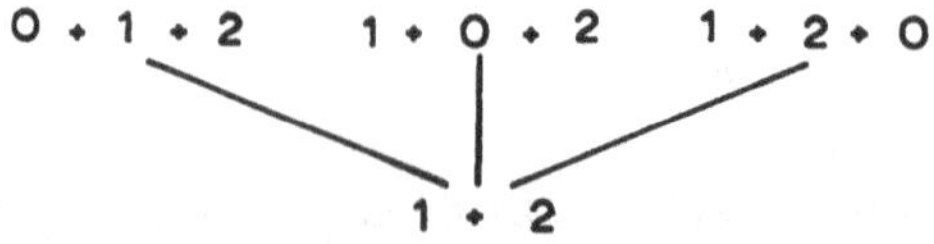

Fig. 10

These are the diagrams that correspond in the general case of any category to:

$$d_1^3(ab) = b \quad d_2^3(ab) = p \quad d_3^3(ab) = a.$$

But now another little problem arises: how is the d^2 to be defined?We could have d_1^2 and d_2^2 operate on words of only one letter. To do so would be to accept simplicial diagrams 'truncated' at the bottom. This is admissable, but we can do better. We need only give names to the unnamed elements of the diagram of Figure 11. Nor shall we find ourselves at a loss for an intuitive interpretation. (This will give us the opportunity to reflect on those popular complexes called 'graphs.)

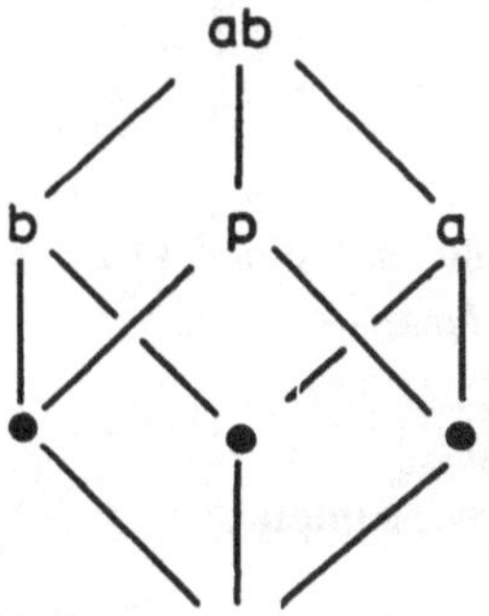

Fig. 11

On the diagram we read:

$$d_2(b) = d_1(a).$$

That is, the arrival (d_1) is identical to the departure (d_2) of b. And similarly, the arrival of b = the arrival of p: $d_1 b = d_1 p$.

But what can be said about f, g and h?They are the *nul* paths and not points. It is here that formal algebra joins our intuition and spontaneous spatial language. Indeed, it is an easy exercise, and we shall show what is to be written (as a result of the equations of 1.2.5):

$$s_1(d_1(a)) = g$$
$$s_1(d_2(a)) = f.$$

The operation thus shows the object 'point' (arrival or departure) at the level 'path'. But it is a nul path.

To have these notations conform to those customarily used in classical geometry (which were for a long time used in simplicial literature), we can rewrite the figure as follows:

$$A \longrightarrow B \longrightarrow C \quad \text{or:} \quad \underline{A \quad B \quad C}$$

and we write

$$f{:}= AA \quad a{:}= AB, \quad g{:}= BB \text{ and so on.}$$

Thus our drawing becomes:

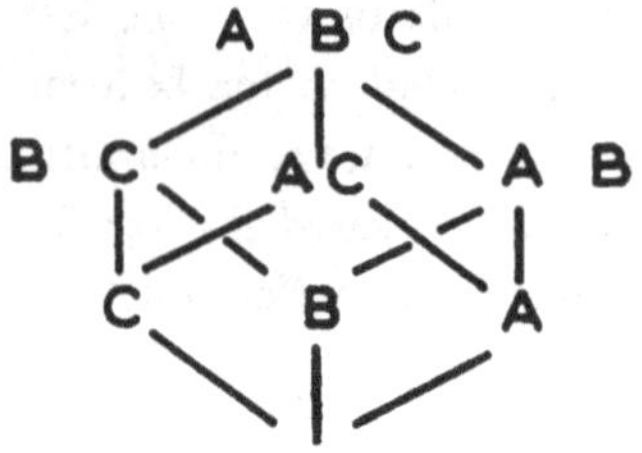

Fig. 12

good old simplex!

Exercise: What plays the role of 'points' in the addition of integers, since the integers are considered to be 'paths'?

3. Simplicial categorys (abstract complex)

When dealing with models which are clearly isomorphic, it is natural, in order to bring out their structure, to adopt a formal point of view and also adopt an abstract definition of what is common to all of the simplexes. We will no longer worry about what the letters *d*, *o*, and *s* mean, no longer ask 'what is it'?, but rather 'how does it work'?

3.1 Formal grammar

3.1.1 The constituents are:

the *o*'s	o^0	o^1	o^2	o^3	...	o^n	...
the *d*'s		d_1^1	d_1^2	d_1^3	...	d_1^n	...
			d_2^2	d_2^3	...	d_2^n	...
				d_3^3	...	d_3^n	...
					...	...	...
						d_n^n	...
the s*'s*		s_1^1	s_1^2	s_1^3	...	s_1^n	...
			s_2^2	s_2^3	...	s_2^n	...
				s_3^3	...	s_3^n	...
					...	...	...
						s_n^n	...

3.1.2 Rules of *associative composition:*

any d^n can be followed by: o^{n-1} or d^{n-1} or s^{n-1}
any s^n can be followed by: o^{n+1} or d^{n+1} or s^{n+1}
any o^n can be followed by: o^n or d^n or s^n

3.1.3 Transformation rules:

$$
\begin{array}{llll}
1) & o^n d_i^n = d_i^n, & o^n s_i^n = s_i^n, & o^n o^n = o^n \\
 & d_i^n o^{n-1} = d_i^n, & s_i^n o^{n+1} = s_i^n & \\
2) & s_i^n d_i^{n+1} = o^n, & s_i^n d_{i+1}^{n+1} = o^n & \\
3) & d_i^n d_j^{n-1} = d_j^n d_{i-1}^{n-1} & \text{if} \quad i > j & \\
4) & s_j^n s_i^{n+1} = s_{i-1}^n s_j^{n+1} & \text{if} \quad i > j & \\
5) & s_i^n d_j^{n+1} = d_j^n s_{i-1}^{n-1} & \text{if} \quad i > j & \\
 & d_i^n s_j^{n-1} = s_j^n d_{i+1}^{n+1} & \text{if} \quad i > j &
\end{array}
$$

(with the usual rules of the = sign, that is, transitivity and substitutability).

The structure formed is a *category* (abstract). We can say: *simplicial category,* but more commonly 'complexes'.

4. Simplicial sets (set complexes)

4.1 The category we have just described (the *d, o, s* and their grammar) can be 'represented' by replacing the *d, o,* and *s* with determined mathematical objects (themselves having an internal structure). For example, set mappings, as was the case for the models presented above (1 and 2). In this case we say that we have a simplicial *set* (a term that can be considered an abbreviation of: the set model of simplicial category) or set *complex.*

Such a model is composed of: 1. the sets $K^0, K^1, K^2, \ldots, K^n, \ldots$ and 2. the mappings:

$$
\begin{array}{llll}
d: & K^0 \longleftarrow & K^1 \leftleftarrows & K^2 \ldots \\
o: & & \updownarrow & \updownarrow \quad \text{etc.} \\
s: & & K^1 \longrightarrow & K^2 \ldots
\end{array}
$$

with the five types of relations given above in (1.2.5) and in (3.1.3).

4.2 The following is another example of a simplex set. Let us choose any set A and form the products $A \times A$, $A \times A \times A$, etc. which will be noted, as usual, A^2, A^3, etc.

There exists, as we know, two mappings

$$A \times A \rightarrow A$$

called projections which will be noted d_1 and d_2.

To the mapping $S_1 : A \rightarrow A \times A$, we will associate the *diagonal* mapping of A which makes, correspondingly a diagonal element (a, a) of A^2.

It is easy to continue in this manner and define three projections, d_1, d_2, d_3 of $A \times A \times A$ on $A \times A$. And two injections which can again be called the diagonals of A^2 and A^3 those to which (a, b), an element of A^2 correspond to (a, a, b) and (a, b, b), elements of A^3, and so on.

It is easy to see that the d and s defined in this manner obey the necessary conditions.

Let us now return to the example given above in (1.3.1), the complex Δ^3 as a subset of the cube $\{\text{P A U L}\}^3$.

4.3 Complexes and simplexes ('hat' operation)

The set K, which is often called a complex (without epithet, as an abbreviation) is composed of the sets K^n and of the mappings:

$$d: K^n \rightarrow K^{n-1} \text{ and the mappings } s: K^n \rightarrow K^{n+1}$$

satisfying the conditions given above (five types of equations).

In the case of a simplex (first model) the elements of the sets were generated from a unique source word.

In the general case, we choose any element $p \in K^m$. To this element we associate a word of m letters, all distinct, call it M. (This is simplicial nomenclature; in the old tradition of classical geometry we would say: 'Given the triangle *ABC.*')

We will now construct a mapping:

$$Ch : \Delta[M] \rightarrow K$$

of the complex type Δ, or a simplex in any complex K. This will be a morphism of the simplicial structure. For this we will write:

$$Ch(M) = p,$$

then

$$Ch(d_iM) = d_ip, \quad Ch(s_jM) = s_jp$$

and so forth.

In this manner we obtain a 'simplicial' nomenclature for a subset of K. That is, the image of Δ by the mapping written *Ch*.
Thus, in every complex K and for every element p, we can construct an image of Δ: it is the ***singular*** simplex attached to p, and which we will write as $\hat{p}$ (hat) Actually it is not, strictly speaking, only the image (subset of K) that is important but the image which has been given the labels that are the words of Δ ( ). Rather, we define the mapping itself as the 'singular simplex'; here provisionally noted '*Ch*'. Finally, we shall write it as:

$$\hat{p} : \Delta[M] \to K$$
$$\hat{p}(p) = M.$$

This explains why we can call those elements of a complex K 'simplexes' (in the set sense), and this is a linguistic simplification that is both classic and useful. We can this represent a complex as an architectural element or structure for which the materials are the simplexes (certain authors use the tilde ∿ instead of the 'hat' ˆ).

5. Products of simplexes

We shall show that every cartesian product of simplexes can be organized as a complex, and this will give us the first process for constructing these complexes.

5.1 In anticipation of what will later be called the (affine) geometrical *construction* of a complex, I will point out that elementary geometry has used simplex structures for centuries, without making them explicit, employing them in the most common affine architectures: points, line segments, triangles, etc.

For example, if I speak of 'the line segment *AB*', I am interested, among other things, in the relation that the points A and B have with the segment in question:

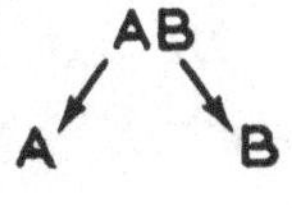

And as I have already pointed out (in 2.4.1 *in fine*) we may have need of 'nul segments' AA and BB.

Therefore, to every segment we will associate a simplex of the type $\Delta(2)$. Let us now consider the cartesian product of two segments; if we adopt the set definition, it will be the set of pairs of points. We say, quite naturally, that the cartesian square of the segment $(0 \leqslant x \leqslant 1)$ is a square (in fact, it will most often be a parallelogram because we are working in affine geometry, which has no right angle).

Below is the traditional image (we say, cartesian), of the product of two segments.

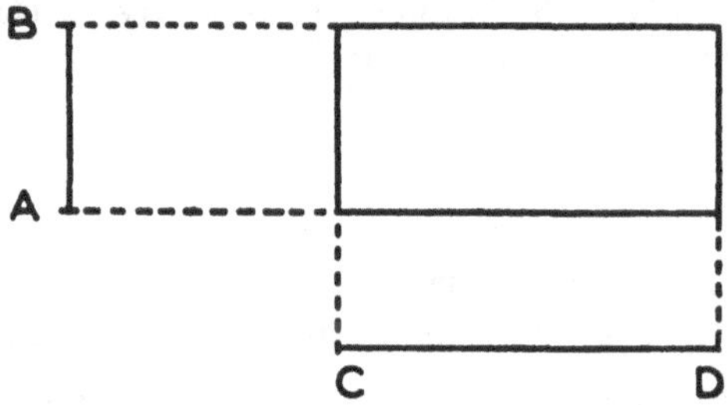

Fig. 13

But the combinatorial architecture of the parallelogram (or the square) is not simplicial: it has four vertices, four sides.

We shall see that this is only apparently so.

5.2 For the moment, let us abandon geometrical images and fix our attention on combinatorial structures. That is, we will consider two simplexes: $\Delta[AB]$ and $\Delta[CD]$.
As we have seen, the first is composed of the sets:

K_1: A, B
K_2: AA AB BB
K_3: AAA, AAB, ABB, BBB

and so on with, in addition, the mappings d and s previously defined.

The second is the same (an isomorph), call it L:

L_1: C, D
L_2: CC, CD, DD
L_3: CCC, CCD, CDD, DDD, etc.

The definition of the product of such structures is obviously:

1. form the products $K_i \times L_i$ for $i = 1, 2, 3, \ldots$; and
2. define the mappings d and s:

The natural way would be:

if $x \in K_i$ and $y \in L_i$, then $(x,y) \in K_i \times L_i$,
and $d(s,y) = (dx,dy)$, $s(x,y) = (sx,sy)$, by definition.

The complex product $M = K \times L$ in our example, will thus be defined by:

M_1: (A, C), (A, D), (B, C), (B, D)
M_2: (AA, CC), (AA, CD), (AA, DD), (AB, CC), ..., (BB, DD)
M_3: ..., (AAB, CCD), (AAB, CDD), ..., (ABB, DDD)

and so on.

We can easily show that the d and the s obey the necessary rules:

$$d^1(AAB, CCD) = (AB, CD)$$
$$d_2(AAB, CDD) = (AB, CD).$$

(The reader should try to draw the appropriate diagram here.)

5.3 The simplex K can be called monogenous in the sense that it is 'generated' by the single word '*AB*', to which we will apply in all possible ways the operations d and s. It is the same for L.

However, this is not quite the same thing for $M = K \times L$. For example, the element (AB, CD) does not suffice to generate M; by using the operation d we only obtain (A, C) and (B, D). No successive applications of d permit us for example, to obtain the element:

$$(AAB, CDD).$$

We can thus ask ourselves how the generator of the elements can be found. I will only give the answer here and the reader should establish for himself that if we begin with:

$$(AAB, CDD) \text{ and } (ABB, CCD)$$

every other element of the M product can be obtained by a sequence of operations of s and d appropriately chosen. That is, all the elements of M can be found in the two simplexes:

$$\Delta[(AAB, CDD)] \text{ and } \Delta[(ABB, CCD)]$$

either in one or the other, or both.

5.4 Let us now return to the geometric interpretation. What corresponds to (*AAB, CDD*) for example?

Performing first all of the *d* operations, we obtain:

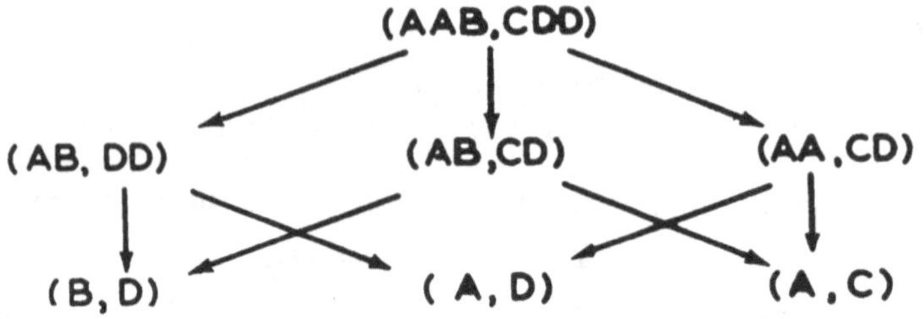

Fig. 14

which illustrates the architecture of the triangle for which the three vertices are: (*B, D*), (*A, D*) and (*A, C*).

The figure can accordingly be drawn:

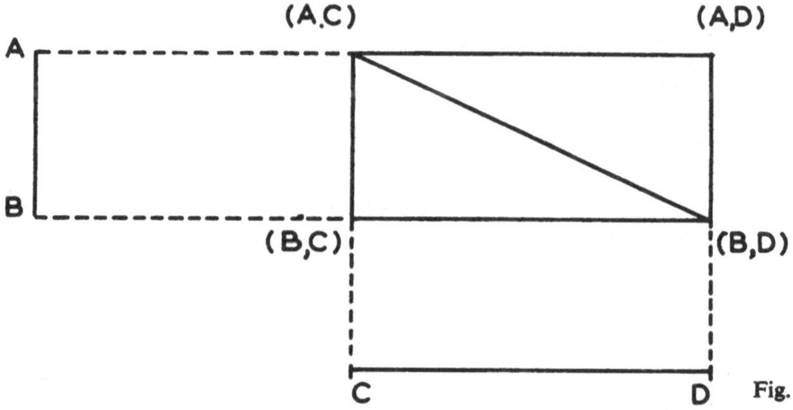

Fig. 15

and the parallelogram is simplicially structured, that is, decomposed into triangles.

5.5 To familiarize ourselves with the mechanism, we can examine two other examples.

5.5.1 The product of a triangle by a segment gives the volume of a triangular prism. We shall examine the complex $\Delta[AB] \times \Delta[CDE]$ and see that it can be generated by the simplexes:

$$\Delta[(AAAB, CDEE)]$$

$$\Delta[(AABB, CDDE)]$$

and

$$\Delta[(ABBB, CCDE)]$$

from whence a decomposition of the prism into three tetrahedrons:
six vertices:

$$(A, C), \quad (A, D), \quad (A, E), \quad (B, C), \quad (B, D), \quad (B, E);$$

twelve edges:

$$\begin{array}{lll} (AA, CD), & (AA, CE), & (AA, DE), \\ (AB, CC), & (AB, DD), & (AB, EE), \\ (AB, CD), & (AB, CE), & (AB, DE), \\ (BB, CD,) & (BB, CE), & (BB, DE). \end{array}$$

5.5.2 The product of three segments is a cube. The complex $\Delta[AB]$, $\Delta[CD]$, $\Delta[EF]$ can be generated from six word-products:

$$\begin{array}{l} (AAAB, CCDD, EFFF) \\ (AAAB, CDDD, EEFF) \\ (AABB, CDDD, EEEF) \\ (AABB, CCCD, EFFF) \\ (ABBB, CCDD, EEEF) \\ (ABBB, CCCD, EEFF) \end{array}$$

which decomposes the cube into six tetrahedra.

Bibliographic notice

1. The paper by Henri Cartan, given in December 1956, entitled 'Sur la theorie de Kan', was published by the *Secrétariat mathématique* of the Henri-Poincaré institute *(Seminaire H. Cartan, E.N.S. 1956–1957).* Reedited in book form by Benjamin Inc. New York 1967.
2. To which must be added the two talks 'which were never given' [sic] for which a resumé appeared in the above mentioned collection the same year under the titles 'Sur le foncteur Hom en théorie simpliciale' and 'Théorie des fibrés simpliciaux' (*ibid,* pp. 3–01, 1–12 and 4–04, 4–12).
3. In the Ergebnisse: *Calculus of fractions and homotopy theory,* by P. Gabriel and M. Zisman (Berlin, Springer, 1967); and particularly in chapter 2: 'Simplicial sets' and chapter 4: 'Homotopic category'.

4. J. Peter May. *Simplicial objects* in *Algebraic topology.* Princeton, Van Nostrand (Van Nostrand Math. Studies, vol. 11), 1967, 161 p.
5. If the reader happens to have the collection cited in (1) above, he can refer to the article by J. L. Moore (1955), *Sémininaire H. Cartan 1954–55,* p. 12–01, and that H. Cartan entitled 'Operations cohomologiques', *ibid,* p. 14–01.
6. Each of the three sources cited contain bibliographic information.

5. monoids and groups

by C. d'Adhémar

Introduction

We will define two types of algebraic structures which are frequently encountered: monoids and groups. These structures have been shown to be useful in various domains such as the monoids in formal linguistics (N. Chomsky and M. Schutzenberger) and groups in diverse branches of the human sciences where they have been used to represent filiation systems, and in particular the rules of kinship and marriage (mathematic research by A. Weil and Ph. Courrege on the work of C. Lévi-Strauss and H. White) and in genetic psychology (J. Piaget). Geomancy, a process of devination, also lends itself to the use of group structure (the Gara studied in the Sara country by R. Jaulin). These mathematical structures are also used in the decorative arts to show symmetries and repetition of motifs (H. Weyl) as well as in the musical domain where the notion of groups can be applied to frequencies (P. Barbaud) and to musical composition (*Nomos alpha* by I. Xenakis, analyzed by F. Vandenbogaerde).

The general theory of these structures will not be presented here, but rather some examples will be given as exercises. The principles used to construct these algebraic structures to be adopted here is as follows: starting with a finite alphabet, we shall write the lexicon of words of finite strings of letters in this alphabet. Then either allowing all words or imposing spelling rules will give us systems that vary, according to the set of rules adopted. Thus, we will be dealing with sets of words to which we will associate compositional laws which confer on them the desired algebraic structures.

We have arbitrarily chosen to begin with an alphabet of two letters, but any finite alphabet could have been used. These structures, built on a finite number of letters, and having a finite or infinite number of elements, are said to be of finite type.

1. Monoids

1.1 Free monoids

1.1.1 Construction of a lexicon

Given two letters $A = \{a, b\}$ let us construct a 'lexicon': the set of all words (including the word containing no letter, or the empty word, denoted here by I) finite in number and without parentheses that can be written with the letters a, b. We shall call this lexicon F. Another way of defining F is:

$$F = I + A + A \times A + A \times A \times A + \ldots + A \times \ldots \times A + \ldots$$

Let us now consider a simple procedure: the construction of a lexicon as a tree. For each word m of $(n-1)$ letters, construct as many branches as there are words of n letters, beginning by m (two branches: adding either a or b to m). There are thus twice as many words of n letters as there are words of $(n-1)$ letters. This tree, all of whose branches double, is the exponential tree to the base 2, otherwise known as a dichotomic tree.

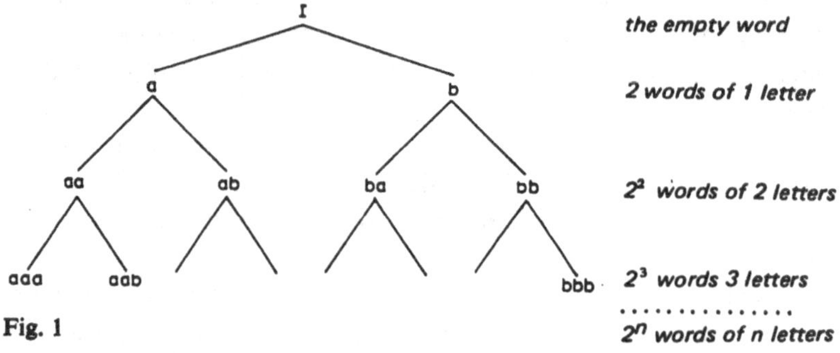

Fig. 1

Exercise: Write all of the words of five letters.

1.1.2 Composition law on the words of the lexicon

The law of internal composition, or binary operation[1] is defined in the lexicon F. It is noted here by the symbol *, an operation by which we obtain the

1. A binary operation T on a set E is a mapping of $E \times E$ into E. In other words, for any ordered pair (x, y) of elements of E there exists one and only one element z of E such that $x\,T\,y = z$.

product of two words m_1 and m_2 by writing the letters of m_2 after those of m_1. For example we have:

$$m_1 = abaa \qquad m_2 = b \qquad m_1 * m_2 = abaab$$

This operation, called a concatenation, is *associative.* If m_1, m_2 and m_3 are words in F, then $(m_1 * m_2) * m_3 = m_1 * (m_2 * m_3)$, since in both cases it is a question of forming a word by writing the letters of m_1, m_2 and m_3 one after the other, respecting their order. On the other hand, for any word m of F, $I * m = m * I = m$. I is called the neutral element for the operation $*$.

A monoid is an algebraic structure consisting of:
1. *a set* M; and
2. *an associative,binary operation defined for any pair of elements of* M *for which there exists a neutral element in* M.

If T is the binary operation in M, the monoid is written (M, T), or simply M, where there is no risk of ambiguity (note that the monoid structure (M, T) is not to be confused with its subadjacent set M).

The lexicon F of finite words constructed from the alphabet $\{a, b\}$, having the operation $*$ is a monoid $(F, *)$ with the neutral element I. No limitation is imposed on the construction of the lexicon; there is no spelling. We shall call such a construction *free.* Thus $(F, *)$ is the *free monoid* generated by $\{a, b\}$, a and b are the generaters of $(F, *)$. In a similar fashion we can construct the free monoid generated by n elements (an alphabet of n letters) with the aid of the exponential tree to the base n, where from each word grow n branches.

1.2 Quotient monoids

We shall now impose spelling rules on the words of F by which we identify words of F that are written differently: certain elements of F are considered as equal. In this manner we obtain a new set F' for which each element is a class of words equal among themselves. This set F', which is given an associative law of composition, derived from $*$ is , like F, a monoid. A few simple examples will serve to illustrate the procedure.

1.2.1 The idempotence rule

This rule can be expressed by: $m * m = m$, for any word m of F. If a word is repeated after itself, it is the same as writing it only once. In particular;

$$aa = a * a = a \qquad bb = b * b = b$$

If one word can be derived from another by iterated applications of this rule, then the two words are considered to be identical. When the number of letters in a word cannot be reduced by applying the idempotence rule we say that it is a *reduced word.* In the case of an alphabet consisting of two letters, it is easy to show that there are seven reduced words. They are:

	I
a	*b*
ab	*ba*
aba	*bab*

Every word is equal to at least one reduced word. In all the examples given here we have chosen rules such that any word will be equal to only one reduced word. In other words, regardless of the method chosen to reduce a word,. we always obtain the same result. If $m = abaaba$ is the given word, it can be reduced in several ways. For example:

1) $m = ab(aa)ba = (ab)(ab)a = aba$;
2) $m = (aba)(aba) = aba$.

But this is not a general result. There exist algebraic structures for which the classes can have several words of minimum length.

The idempotence rule divides the free monoid F into seven disjoint classes (or partitions F into seven classes) each containing a reduced word to which all the words in each class are equal. In particular, I here is the only member of its class. For example the class *ab* contains *abab, abaab,* etc. . . .

We must now give this set of classes F' an operation $\overline{*}$ which will transform it into a monoid.

If C_1 and C_2 are classes, $C_1 \overline{*} C_2$ is the class of reduced words which is equal to the product of any two elements of C_1 and C_2. This procedure is possible because the product by $*$ of two elements depends only on the reduced elements to which they are equal (that is, of their classes). For example:
if:

m_1 belongs to the class of the reduced word *aba*, and
m_2 belongs to the class of the reduced word *ab*,

then,

$$m_1 \overline{*} m_2 = aba\ ab = ab.$$

It is easy to show that this operation is associative. Thus, we have obtained a monoid F', called the *quotient monoid* of F (that is, we have obtained a set of classes of a partition of F) having a composition law derived from that of F.

This procedure of partitioning a set in such a fashion allows us to give the quotient an operation, derived from the first and having the same properties; this is the general process of constructing *quotient structures.*

Instead of considering the set of classes, it is also possible to consider the set L of reduced words having the operation $*'$: the product by $*'$ of two reduced words is the reduced word equal to the product obtained from the operation $*$ of these two words. For example:

$$aba *' ba = aba$$

The monoid L composed of these seven words is called *Kuratowski's monoid.*

Sometimes there is no distinction made between L and F'; they are isomorphic (see section 3). Nevertheless, in order to be truly rigorous it is desirable to make the distinction between seven elements and seven classes.

The Pythagorean table of the $*'$ operation for Kuratowski's monoid can be constructed in the same way as the addition tables or the multiplication tables with which we are familiar. The table can also be used as a table for $(F' \overline{*})$. Each reduced word serves as a label or representative of its class.

	I	*a*	*b*	*ab*	*ba*	*aba*	*bab*
I	*I*	*a*	*b*	*ab*	*ba*	*aba*	*bab*
a	*a*	*a*	*ab*	*ab*	*aba*	*aba*	*ab*
b	*b*	*ba*	*b*	*bab*	*ba*	*ba*	*bab*
ab	*ab*	*aba*	*ab*	*ab*	*aba*	*aba*	*ab*
ba	*ba*	*ba*	*bab*	*bab*	*ba*	*ba*	*bab*
aba	*aba*	*aba*	*ab*	*ab*	*aba*	*aba*	*ab*
bab	*bab*	*ba*	*bab*	*bab*	*ba*	*ba*	*bab*

We have thus constructed a monoid having seven elements (of order seven[2]) from a monoid having an infinite number of elements (of infinite order[3]).

2. For the case where we have a finite number of generators and rules, we do not always know how to determine if the set constructed has an infinite number of elements, or none. This is an old problem, known as Burnside's problem.

3. The order of a monoid is the number of its elements (as is the case for any algebraic structure).

1.2.2 Commutative rule

The order in which the words are written is not significant: $m_1 * m_2 = m_2 * m_1$ for any words m_1 and m_2 in F. In particular, $ab = ba$, and the word is not changed by changing the order of the letters. As a result, two words are equal if, and only if, they have the same number of a and the same number of b. We reduce a word, for example, by putting it in the form $a^n b^q$ if it contains n times the letter a and q times the letter b:

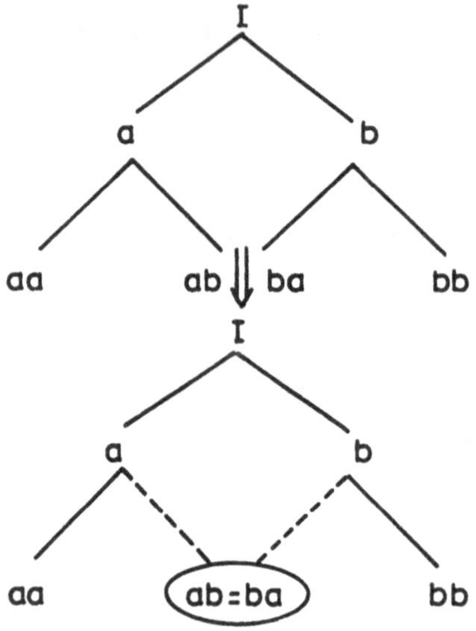

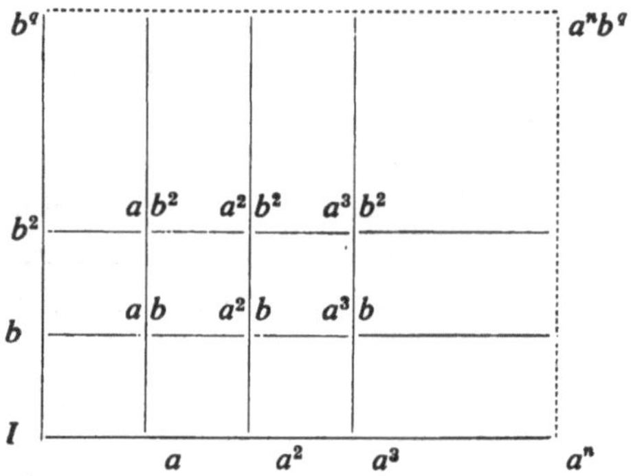

Fig. 2

From the tree (Fig 4) we can construct the network which can be represented by the network N^2 of whole pairs:

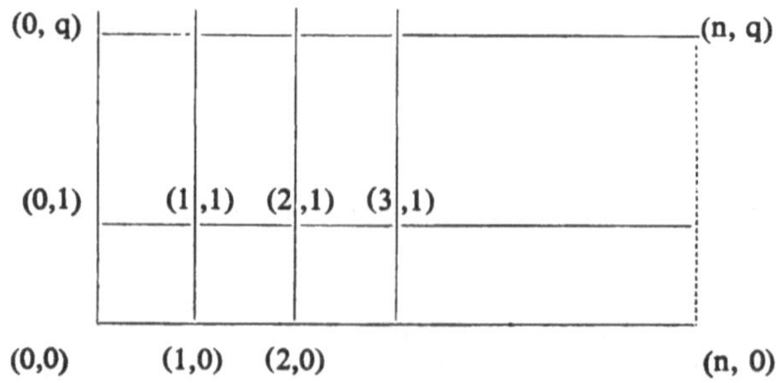

As was the case for the idempotence rule, the commutative rule divides F into disjoint classes containing a reduced word and the words equal to it.

This set F'' of classes (or the set L' of reduced words) can be assigned an associative operation $\overline{*}'$ derived from $*$:

$$a^n b^q \;\overline{*}'\; a^{n'} b^{q'} = a^{n+n'} * b^{q+q'} = a^{n+n'} b^{q+q'}$$

F'' and L' are quotient monoids having an infinity of elements (see figure): there are $(n+1)$ reduced words of n letters for any integer n. In Figure 3, the words of n letters are located on the nth diagonal.

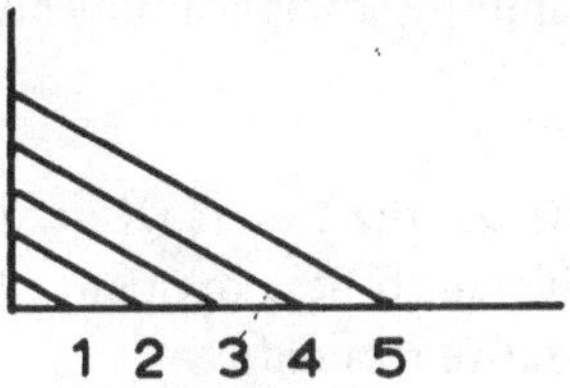

Fig. 3

1.2.3 Idempotence and commutativity

We have succesively imposed idempotence and commutativity. What happens when we require both? For L (idempotent) we have:

$$aba = aab = ab;\; bab = abb = ab \text{ and } ba = ab.$$

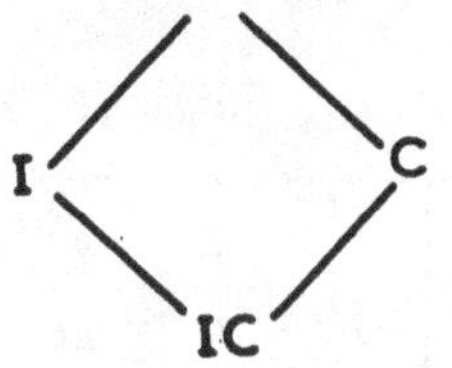

Fig. 4

Thus, L is divided into four classes: $\{I\}$, $\{a\}$, $\{b\}$, $\{ab, aba, bab, ba\}$. The quotient monoid L'' can be derived from the Pythagorean table as follows:

	I	*a*	*b*	*ba*	*ab*	*aba*	*bab*
I							
a						*ab*	
b							
ba							
ab		*ab*				*ab*	
aba							
bab							

	I	*a*	*b*	*ab*
I	*I*	*a*	*b*	*ab*
a	*a*	*a*	*ab*	*ab*
b	*b*	*ab*	*b*	*ab*
ab	*ab*	*ab*	*ab*	*ab*

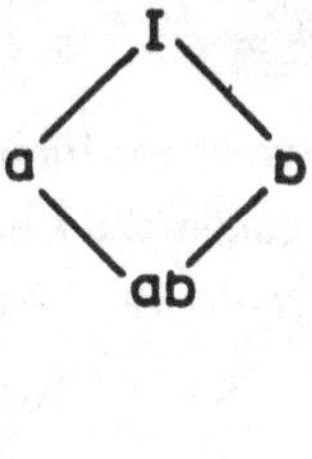

Fig. 5

The elements of L'' correspond to the four subsets of $\{a, b\}$. Indeed, two words are distinct only if they do not contain the same letters; the order of the letters is immaterial (commutativity), as is the number of identical letters in the word (commutativity and idempotence). More generally, with an alphabet of n letters we will again have a bijection between the reduced words and the subsets of the alphabet. We find here the simplexes S_n (subsets of a set of n elements and the inclusion relation). (See chapter 4, 'Simplical objects', and chapter 2, 'Ordering and classification'.)

Exercise: Show that L'' the quotient of L (Kuratowski monoid having seven elements) can also be considered to be a quotient of F a free monoid generated by $\{a, b\}$ and of L' (commutative monoid).

1.2.4. Commutativity and $a^2 = b^2 = I$

We shall see in 2.2 that we have a group: Klein's group. We obtain the same letters as above, but the table is different.

	I	a	b	ab
I	I	a	b	ab
a	a	I	ab	b
b	b	ab	I	a
ab	ab	b	a	I

2. Groups

We have just seen an example (1.2.4) where for any element m there exists one, and only one, element which we will write as m^{-1}, such that $m^{-1}m = I$ (as can be seen from the preceding table). Similarly for (1.2.4):

$$a^{-1} = a \qquad b^{-1} = b \qquad (ab)^{-1} = ab$$

and for any word m: $Im = mI = m$.

I is thus called the *neutral element*, and m^{-1} the *inverse* of m for the binary operation. Such a set is a group. To be more precise:

A group is an algebraic structure composed of:

A set E;

a binary operation ○ *defined for every ordered pair of elements of* E; *and:*

the operation ○ *is associative,*

E *contains a neutral element for the operation.*

every element of E *has an inverse.*

A group will be written $(E, \bigcirc)$ or simply as E where there is no risk of ambiguity (but to be rigorous, E is only the subjacent set of $(E, \bigcirc)$. A group can also be defined as a monoid for which every element has an inverse or an *invertible monoid.*

Klein's group (1.2.4) is a well-known group. It is a group of four elements characterized by the following: each element is its own inverse; and the product of two elements different from the neutral element is equal to the third.

2.1 Free groups

As is the case for the monoids, we can construct the free group G generated by $\{a, b\}$. To do so, we add the 'letters' a^{-1} and b^{-1}, the inverses of a and b, and we can construct the free monoid generated by $\{a, b, a^{-1}, b^{-1}\}$ to which we assign the spelling rules:

$$a^{-1} a = a a^{-1} = b^{-1} b = b b^{-1} = I$$

The set of words that have been reduced by successive deletions of two inverse letter written side by side constitutes the free group generated by $\{a, b\}$. For this to hold, it is necessary that every word in the set be equal to one, and only one, reduced word and nothing else. For a detailed proof see Marshall Hall (1959, chapter 7).

Each reduced word m has one, and only one, inverse m^{-1} obtained by reversing the order of the letters and the sign of the power:

$$m = a b a b^{-1} b \qquad m^{-1} = b^{-1} b a^{-1} b^{-1} a^{-1}$$

$$mm^{-1} = a (b (a (b^{-1} (b b^{-1}) b) a^{-1}) b^{-1}) a^{-1} = I$$

To construct the tree of this group it is sufficient to prune the tree of the monoid generated by $\{a, b, a^{-1}, b^{-1}\}$ of the branches which add the inverse letter of the last letter of the word: each word will thus give rise to only three branches, except for the empty word.

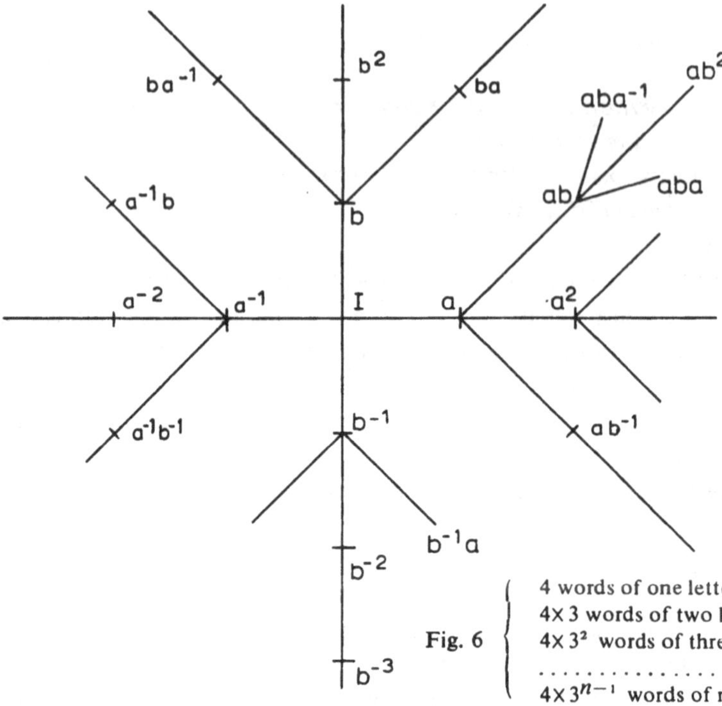

Fig. 6

4 words of one letter
4×3 words of two letters
4×3^2 words of three letters
. . . .
$4 \times 3^{n-1}$ words of n letters

2.2 Quotient groups

We shall now add rules which allow us to obtain disjoint classes and an associative operation on these classes, derived from $*$ (thus showing that the quotient set is indeed a group for which the neutral element is the class of the neutral element of G, and such that the inverse of a class f is the class of the inverse of any element of f).

Idempotence: Any idempotent group can be reduced to the neutral element. Indeed, if $x = xx$, $I = x^{-1}\ x = x^{-1}(xx) = x$.

Commutativity: The quotient is a commutative group, or abelian, (see chapter 6, 'Measure scales'). Each class can be represented by its unique reduced word of the form $a^p\ b^q$, where p and q are positive or negative integers.
Commutativity and $a^2 = b^2 = I$:
We find the Klein group already encountered in 1.2.4.
Commutativity and $a^3 = b^2 = I$:

The reduced words:

$$I$$
$$a \qquad b$$
$$a^2$$
$$ab$$
$$a^2b$$

form a group G' of six elements (group of order six) for which the table is:

	I	a	a^2	b	ab	a^2b
I	I	a	a^2	b	ab	a^2b
a	a	a^2	I	ab	a^2b	b
a^2	a^2	I	a	a^2b	b	ab
b	b	ab	a^2b	I	a	a^2
ab	ab	a^2b	b	a	a^2	I
a^2b	a^2b	b	ab	a^2	I	a

Let us examine this example. We note that in this table, each element of G' appears once, and only once, on each line and in each column. In other words, the equation $\alpha x = \beta$ has but one solution in x when α and β are given. That is: $x = \alpha^{-1}\beta$, and similarly, the equation $y\alpha = \beta$ has only one solution: $y = \beta\alpha^{-1}$. This holds for every group.

On the other hand, the set $\{I, a, a^2\}$ (top left entry of the above table) is itself a group generated by a single element (which we call a monogenous group) and is such that $a^3 = I$ is a cyclic group.

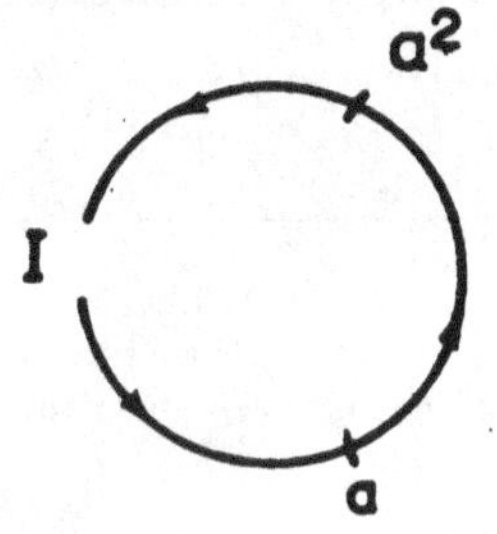

Fig. 7

Here the cyclic group is of order three which can be represented by interpreting a as a third turn rotation (do not confuse rotations with image points. Here, for example, the rotation $2\pi/3$ transforms the image point I into a, the image point a into a^2, and the image point a^2 into I).
A subset of G which is itself a group for the composition law of G is called a subgroup of G.[4] $\{I, a, a^2\}$ is thus a subgroup of G'.

If we partition G' into two classes indicated by the thick lines of the table: class A is composed of the elements of the cyclic group $\{I, a, a^2\}$, and class B is composed of the words written with b ($B = bA$). We note that these two classes form a group G'', quotient of G', of neutral element A:

	A	*B*
A	*A*	*B*
B	*B*	*A*

To every subgroup of a commutative group can be associated a quotient group whose neutral element is the class constituted by this subgroup (called a kernel), every other class (Lagrange classes) being formed by the products of an element of the group by this subgroup (all of those classes having the same number of elements). Here, this kernel subgroup, the neutral element of the quotient group, is made of words considered to be equal to the empty word according to the rules imposed. Show that they do constitute a subgroup.

	I	b	a	ab	a^2	a^2b
I	I	b	a	ab	a^2	a^2b
b	b	I	ab	a	a^2b	a^2
a	a	ab	a^2	a^2b	I	b
ab	ab	a	a^2b	a^2	b	I
a^2	a^2	a^2b	I	b	a	ab
a^2b	a^2b	a^2	b	I	ab	a

	A'	A''	A'''
A'	A'	A''	A'''
A''	A''	A'''	A'
A'''	A'''	A'	A''

4. Generally speaking, if S is an algebraic structure, we will call a substructure of S a subset of the basic set of S, which, having been given the laws and relations of S, has the same structure as S.

By forming the quotient of G' and the subgroup $A = \{I, a\, a^2\}$ we obtain the cyclic group of order two (G'').

It is also possible to form the quotient of G' by the subgroup $A' = \{I, b\}$ which is cyclic of order two. The quotient is thus a cyclic group of order three $\{A', A'', A'''\}$. The classes are: A', $aA' = A''$, $a^2A' = A'''$.

Inverse exercise: It is possible to obtain G' by forming the 'direct product' from a cyclic group of order two and a cyclic group of order three:
let G_1 be the cyclic group = $\{I, b\}$
G_2 be the cyclic group = $\{I, a, a^2\}$.
From these two groups we can construct a third, the *direct product group* $G_1 X G_2$ obtained by assigning to the product set (the set of ordered pairs (x, y) where $x \in G_1$ and $y \in G_2$) the operation:

$$(x_1, y_1) \cdot (x_2, y_2) = (x_1\, x_2, y_1\, y_2)$$

where $x_1\, x_2$ is the product in G_1 of the elements x_1 and x_2; and
$y_1\, y_2$ is the product in G_2 of the elements y_1 and y_2.
We thus do obtain the group G' if we agree to simplify the pair (a, b) by writing ab. For example:

$$(a, b) \cdot (a^2, I) = (aa^2, b\, I) = (a^3, b) = (I, b) \rightarrow b$$

Thus, G' has been obtained as the direct product of G_1 and G_2. Similarly, the direct product of two cyclic groups of order two, $\{I, a\}$ and $\{I, b\}$ is the Klein group seen earlier.

3. Homomorphisms

The structure of a quotient group is derived from that of the group which was used to construct it. This can be expressed by the presence of homomorphism, or mapping, that preserves these structures.

A homomorphism of a group H *into a group* H′ *is the mapping* f *of* H *into* H′ *such that* $f(x \cdot y) = f(x) \top f(y)$ *where* $\cdot$ *is the binary operation of* H *and* T *that of* H′.

Let f be the mapping which to each element of the group G' of the preceding example associates its class A or B:

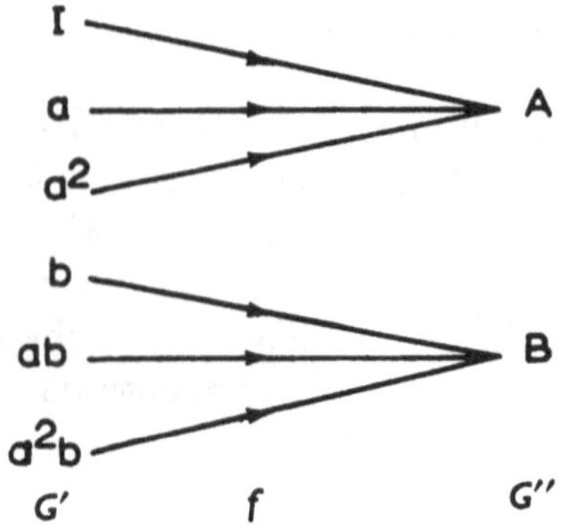

Fig. 8

This mapping is a *homomorphism of the group* G' into the group G'' because the product of the images is equal to the image of the products.

Example: $f(a) = A \qquad f(a^2b) = B$

$$f[a(a^2b)] = f[b] = B = AB = f(a)f(a^2b)$$

A group and a quotient group of this group are always homomorphic. Conversely, we can construct a quotient group of G from any homomorphism f of a group G into a group G' by taking as the nul element of the quotient the set N of the elements of G which have for their images by f the nul element of G'. N is the kernel of f. (Show that we are really dealing with a subgroup of G.) In the case we saw above, where, starting with a free group G, we constructed a quotient group G' by adjunction of rules for which the classes contained one, and only one, reduced word we saw that these reduced words also formed a group G''. The mapping which associates its only reduced word to one class is a bijective homomorphism of G' into G''. We use the term *isomorphism* for this mapping.

Similarly, we can define an *endomorphism* as a homomorphism of a group into itself. For example, if G is the free group generated by $\{a, b\}$, the mapping of G into G, which to each word makes correspond the word obtained by deleting all the b and b^{-1}, is an endomorphism of G:

$$f(aba^{-1}ab) = aa^{-1}a = a$$

The endomorphism of G, defined by the exchange of the letters a and b: $g(a) = b$, $g(b) = a$ (so that $g(a^2ba^{-3}b^2) = b^2ab^{-3}a^2$), is an isomorphism of G into itself; in other words, an *automorphism of G.*

To introduce groups it would have been possible to begin with the general definition and then give the classical examples of sets satisfying the properties of groups (the set Z of integers with the addition operation; the set R of reals with the operations of addition or multiplication). The reason for introducing groups from the point of view of free groups is that there exists a theorem (not given here) that shows that every group is isomorphic to the quotient of a free group.

In other words, any group can be obtained to within an isomorphism by constructing a free group which has been given the appropriate rules. We have seen futher a procedure for the construction of new groups from given groups by forming their direct product groups. Using finite groups of order p and q, it is thus possible to construct a group of order pq.

We shall now turn our attention to a way of representing any group: by permutations.

4. Permutations

A *permutation* on a set E is a bijective mapping of E into E. If $E = \{a, b, c\}$, then the mapping f defined by $f(a) = b$, $f(b) = a$, $f(c) = c$ is a permutation and we represent it as:

$a\ b\ c$
$b\ a\ c$

or by the diagram

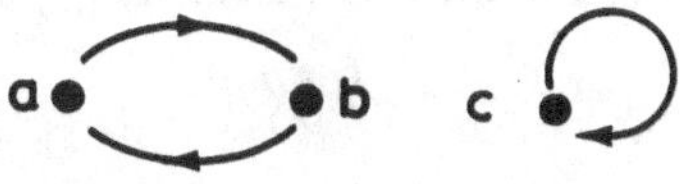

Fig. 9

or by simply abridging the notation by writing the order $b\ a\ c$. This last notation is a consequence of the fact that there are as many permutations as there are ways of arranging the elements of E (but be careful not to confuse arrangements and the permutations that correspond to them).

For a set having two elements $F = \{a, b\}$, there are two permutations. (Indeed, we can take as the first element of the order either a or b and the choice of the first dictatus the second.)

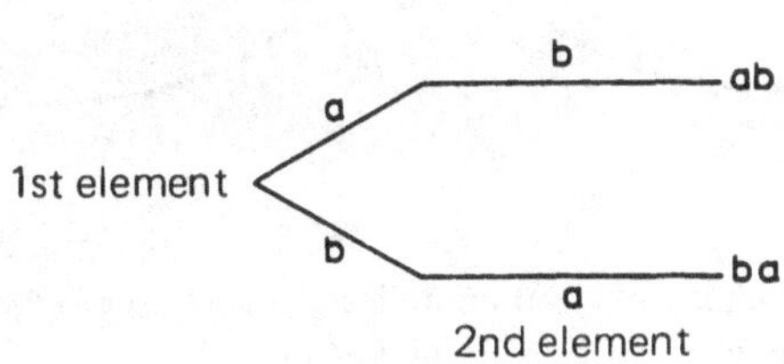

Fig. 10

For $E = \{a, b, c\}$, there are three choices possible for the first element, and having made the first choice, there are two choices left, then one.

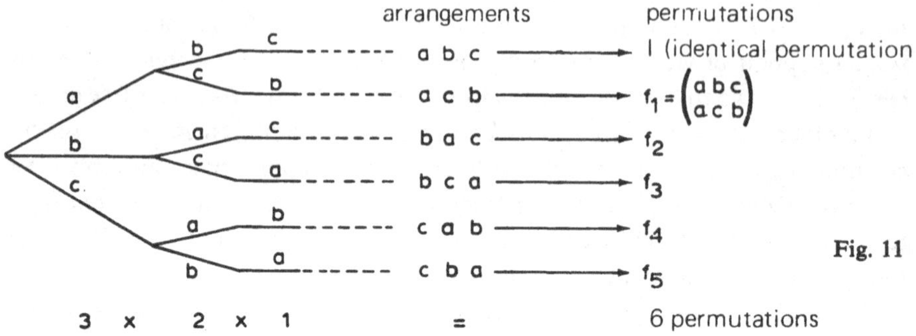

Fig. 11

By definition $n!$ (n factorial) is the number $1 \times 2 \times 3 \times \ldots \times n$. If we write P_n as the set of the permutations of a set of n elements, P_n has $n!$ elements. We have written $P_3 = \{I, f_1, f_2, f_3, f_4, f_5\}$ as the set of the six permutations on E. I, which associates each element to itself, is the identity mapping. P_3 can as usual be assigned the usual operation of composition mapping[5]:

$$f_i \, o \, f_j, \ (i, j) \in \{1, 2, \ldots, 5\}$$

is the mapping obtained by performing f_j and then:

$$f_i : f_i \ o \ f_j(x) = f_i(f_j(x)).$$

It is easy to show that the results of two permutations is a permutation and that the operation o is associative.

On the other hand, $f_i \, o \, I = I \, o \, f_i = f_i$ for any $f_i \in P_3$.

In P_3 each element f has an inverse f^{-1} obtained by taking as the image by f^{-1} of an element, the element for which it is the image. Or, alternatively, we can obtain f^{-1} by reversing the direction of the arrows of f:

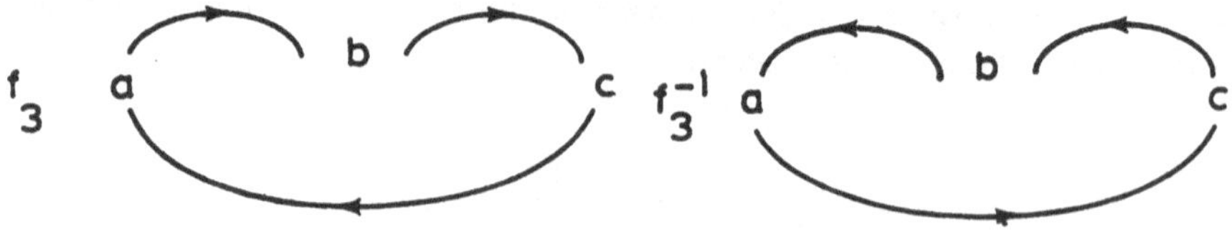

Fig. 12

5. Let f be a mapping of a set A into a set B, and g a mapping of B into a set C. The composition $g \, o \, f$ is the mapping of A into C defined by $g \, o \, f(x) = g(f(x))$ for every element x of A. [$g \, o \, f(x)$ is the image by g of the image by f of x]

P_3 is thus a group.

The reader is advised to construct the Pythagorean table; he can thus verify that in particular this group *is not commutative,* contrary to all the previous examples. For example (by referring to the diagram);

$$f_1 \circ f_3 = \begin{pmatrix} abc \\ cba \end{pmatrix} = f_5 \qquad f_3 \circ f_1 = \begin{pmatrix} abc \\ bac \end{pmatrix} = f_2$$

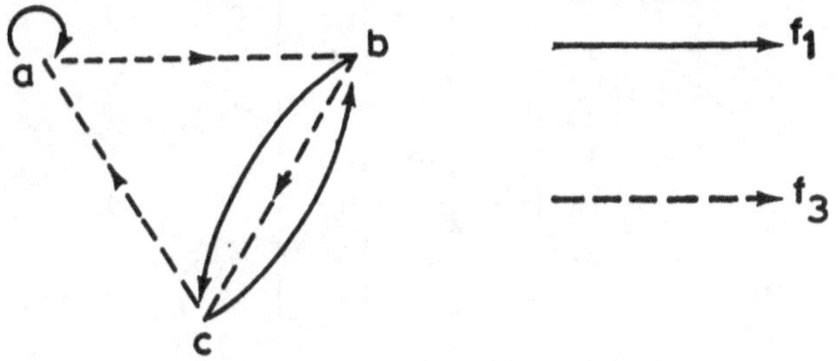

Fig. 13

For any integer n, P_n is a group called the *symmetric group of degree n*. The importance of these groups of permutations is due to the fact that any group G, finite or not, is isomorphic with a subgroup of the group of permutations of the set of the elements of G.

To each element a of G we can associate a mapping t_a of G into G, called the left translation of G; to any $x \in G$ we associate $t_a(x) = ax$. This translation is a permutation: every element y in G is the image of a^{-1}, and is the image of only this element. Indeed, if z is two times the image, that is, if ax_1 $ax_2 = z$, then $a^{-1}ax_1 = a^{-1}ax_2$ and $x_1 = x_2$.

This set T_G of translations the subset of the set of permutations is, when assigned o, a subgroup of P_G. Indeed, by virtue of the associativity in G we have:

$$t_a \circ t_b = t_{ab}$$

since

$$(t_a \circ t_b)(x) = t_a(t_b(x)) = t_a(bx) = a(bx) = (ab)x = t_{ab}(x).$$

The identity mapping is the translation t_e, if e is the neutral element (identity) of G, and similarly $(t_a)^{-1} = t_{a^{-1}}$.

In this manner the mapping f of G into T_G, defined for any $a \in G$, is indeed an isomorphism of the group G into the group T_G.

From this follows the well-known *Cayley theorem:*

Every group is isomorphic to a subgroup of the group of permutations on the set of its elements.

Thus, we can represent any group as a subgroup of the group of permutations on a set.

Example: the Klein group already given (the notations are not the same as those used above):

	e	x_1	x_2	x_3
e	e	x_1	x_2	x_3
x_1	x_1	e	x_3	x_2
x_2	x_2	x_3	e	x_1
x_3	x_3	x_2	x_1	e

can be interpreted as a subgroup of the permutations of the set of its four elements (e, x_1, x_2, x_3):

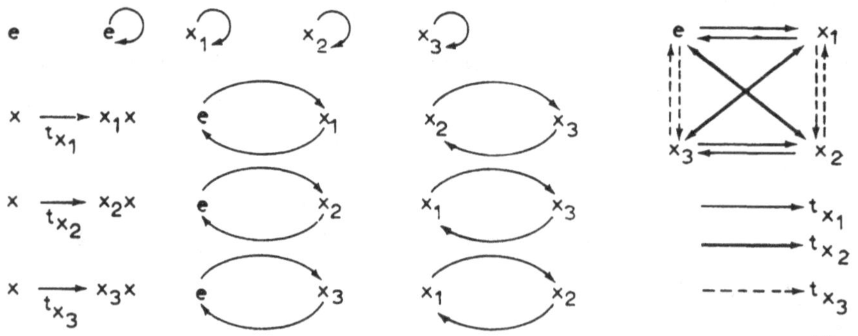

Fig. 14

Another example: cyclic group of order two:

	e	x
e	e	x
x	x	e

which can be interpreted as the group of permutations on the set of its two elements:

Fig. 15

To recapitulate: we have presented a few monoids and groups which are but a small sample of these structures for which we will now give the catalogue:

– a free monoid having an infinite number of elements, generated by $\{a, b\}$ . . . . (Figure 1, p. 128)
– a quotient monoid having an infinite number of elements . . . . (Figure 2, p. 132)
– a monoid of seven elements . . . . (Table, p. 131)
– a monoid of four elements . . . . (Figure 5, p. 133)
– a free group having an infinite number of elements, generated by $\{a, b\}$ . . . . (Figure 6, p. 136)
– a quotient group having an infinite number of elements . (Section 2.2, p. 136)
– a group of six elements . . . . (Table, p. 137)
– a group of four elements . . . . (Table, p. 134)
– a group of two elements . . . . (Table, p. 138)

Bibliography

1. Mathematics

Introductory works:

Barbut, M., 'Le sens du mot *structure* en mathématiques'. *Les Temps Modernes* (Paris), November 1966.

Ledermann, W., *Introduction to the theory of finite groups.* London, Oliver and Boyd, 1953.

Lentin, A. et Gross, M., *Notions sur les grammaires formelles.* Paris, Gauthier-Villars, 1967.

Papy, G., *Groupes.* Paris, Dunod, 1967.

Trahtenbrot, B. A., *Algorithmes et machines à calculer.* Paris, Dunod, 1963.

More technical works:

Coxeter, H. S. M., and Moser, W. O., *Generators and relations for discrete groups.* New York, Springer, 1965.

Magnus, W., Karrass, A., and Solitar, D., *Combinatorial group theory.* New York-London, John Wiley, 1966.

Marshall Hall, Jr., *Theory of groups.* New York, MacMillan, 1959.

Saaty, T. L., *Lectures on modern mathematics.* New York-London, John Wiley, vol. II, 1964 (see Marshall Hall Jr., 'Generators and relations in groups: the Burnside problem').

2. Applications

Barbaud, P., *La musique, discipline scientifique.* Paris, Dunod, 1968.

Chomsky, N., and Miller, G. A., *L'Analyse formelle des langues naturelles.* Paris-La Haye, Mouton/Gauthier-Villars, 1968.

Courrège, P., 'Un modèle mathématique des structures élémentaires de parenté', *L'Homme* (Paris-La Haye, Mouton), juillet-décembre 1965.

Jaulin, R., *La géomancie, analyse formelle.* Paris-La Haye, Mouton (*Cahiers de l'Homme*), 1967.

Kemeny, S., Snell. S. and Thompson, G., *Introduction to finite mathematics.* Englewood Cliffs, N. J., Prentice Hall, 1957, chap. VII.

Lévi-Strauss, C., *Les structures élémentaires de la parenté.* Paris, P.U.F., 1949[2]; Paris-La Haye, Mouton, 1967, 1971[3] (see Chap. 14 'Appendice à la première partie: Sur l'étude algébrique de certains types de lois de mariage (Système Murngin)', by André Weil, pp. 257–265).

Piaget, J., Fraisse, P., and Reuchlin, M., *Traité de psychologie expérimentale.* Paris, P.U.F., 1967, vol. VII.

– and Inhelder, P., *De la logique de l'enfant à la logique de l'adolescent.* Paris, P.U.F., 1955.

Vandenbogaerde, F., 'Analyse de *Nomos alpha* de Iannis Xenakis'. *Mathématiques et Sciences Humaines, Humaines,* n° 24, 1968.

Weyl, H., *Symétrie et mathématique moderne.* Paris, Flammarion, 1964.

White, H., *An anatomy of kinship.* Englewood Cliffs, N. J., Prentice Hall, 1963.

Xenakis, I., *Musiques formelles.* Paris, Richard-Massé, 1963.

6. measure scales

totally ordered abelian groups

by M. Barbut

Introduction

The elementary teaching of the methods of observation of facts and the analysis of observed data in the social sciences are dependent on measure scales, the classic gradation in nominal scales, ordinal or qualitative scales, and numeric or quantitative scales. This distinction is necessary, in fact, since from one class to another the combinatoric and algebraic possibilities, hence the calculations are increasingly expanding.

A nominal scale is a set having no structure, either in the form of relationships, or in the form of operations between its elements. In this respect, it is useful to point out that in the expression 'nominal scale' the adjective 'nominal' is very well chosen. (From a certain point of view, to form a finite collection into a set, in the mathematical sense of the term, is to be able to name each of its elements.) However, the term 'scale' is less satisfactory; 'scale' evokes the notion of arranging the objects in a certain order: into that of ranks. This implies an order structure. We shall not discuss nominal scales here because this topic comes under the heading of sets and classification (partitions) treated elsewhere in this publication (see chapter 2 'Ordering and classification').

Ordinal scales (the term 'ordinal' is preferable because it is more precise than 'qualitative') are those that permit the comparison from the largest to the smallest, from the best to the worst, and in general, from the most to the least. With respect to applications, the fields covered by ordinal scales are those of the scales of sensation, of attitudes, of preference, or utility, etc. They are sets that have been given an order *relation*, in the sense that has been defined elsewhere (see chapter 2) and even a *total* order since two 'ranks' or 'degrees' of a scale are always comparable: one comes before the other. Nevertheless, the last word has not been said on the subject of ordinal scales; their class includes a large variety of types as can be seen immediately by the difference in the nature of categories. On the one hand there are *finite scales*, that is, those that contain a finite number of ranks, from the first (the smallest) to the last (the largest) as do all the instruments of measurement. On the other hand, there is the concept of 'continuum' so often used in the social sciences. When we use this word we generally think of an order (the points on a straight line, for example) such that it contains an *infinity* of elements, and furthermore such that between any two elements we can always insert a third, so that the notion of a scale – in which each degree is preceded and followed by another – makes no sense. We shall have to go further into this typology of ordinal scales (see section 1).

As for numerical scales, (here too, 'numerical' seems preferable to 'quantitative') that is, those scales whose elements are numbers, they permit not only the comparison but also the combination between their elements through the use of *operations*: addition and subtraction, which go together; multiplication and division, etc. The numbers serve to *compare* and to *calculate*. A numerical scale has a double structure: it is totally orderable (comparable) and additive (calculable), if we choose to restrict ourselves to addition as the operation. Moreover, these two structures agree well with each other (in the sense that will be made precise later (section 3), in that the numerical scales can first be classed according to the type of their order. But this classification is insufficient: for the same type of order there can be different possibilities of calculation. It should be understood that one can always perform additions and subtractions on the elements of the scale, but there are cases where one can *divide* the elements in such a way that we have an arbitrary number of equal parts of each of them, and there are cases where we cannot. We know that this possibility of division plays a fundamental role in questions of measure, and more precisely in *commensurability*. A typology of algebraic structures characterized by an operation with the properties common to all additive operation, that is a typology of *abelian groups*, must be outlined, therefore, before we will be in a position to see how such operations can be put in accord with an order structure. This will permit us to more easily analyse what, in the properties of numerical scales, is relevant to the rules to calculus; and how these two components – order and calculation – can later be combined.

It is true that the methodologies of the social sciences generally class the numerical scales from an entirely different point of view and refer to additive scales, multiplicative scales, logarithmic scales, interval scales, etc. This point of view is much less significant, and without doubt, less profound than that which is traditional in mathematics, which we shall use here. As we shall see (see sections 4 and 5) it is only a question of *presenting* objects of the same and only class of mathematical objects: totally ordered, Archimedean, abelian groups.

In the outline to follow of the typology of order and of the additive calculus, we shall retain (because of space limitations) only that which is essential to our subject: the measure scales. The works cited in the bibliography will introduce the reader to more advanced study.

1. Elements of the typology of ordinal scales (types of order)

An ordinal scale (we shall simply call it a 'scale') is a total ordering, that is, a set (or domain of the scale) to which has been assigned a binary relation, noted $\geqslant$, between any two elements. The relation $\geqslant$ must have the following four properties: for any elements a, b and c of the set E,

1. $a \geqslant b$ or $b \geqslant a$ (totality of $\geqslant$)
2. if $a \geqslant b$ and $b \geqslant c$, then $a \geqslant c$ (transitivity of $\geqslant$)
3. if $a \geqslant b$ and $b \geqslant a$, then $a = b$ (antisymmetry of $\geqslant$)
4. $a \geqslant a$ (reflexivity of $\geqslant$).

In the current use of the comparative, which the total ordering formalizes, reflexivity is not generally used. A length, for example, is not greater than itself. Also, this axiomatic property of total order is secondary and we will speak here of a strict order if the relation (total, transitive and antisymmetric) is not defined except between distinct elements of the scale: thus, it will be noted $>$.

Among the various types of scales, we shall first consider the finite scales: those which have only a finite number of degrees:

$$1\text{st} < 2\text{nd} < 3\text{rd} < \ldots < 12\text{th},$$

if there are 12 degrees. These are the only scales that are 'natural', concrete: every instrument of measurement is a finite scale. But the finite scales immediately lead to infinite scales in which the succession of degrees is *unlimited*, so that every finite scale, however large the number of degrees it may have, is nevertheless a subset of an infinite scale. The prototype is the unlimited succession of ordinal integers:

$$1\text{st} < 2\text{nd} < 3\text{rd} < \ldots < n\text{th} < \ldots$$

G. Cantor, the principal author of the typology of order during the years 1880 to 1900, named this *type* ω (omega). It is a type characterized by the fact that there is a first (and smallest) element of the scale, but not a last (greatest) element, that every element has one, and only one, successor – distinct from itself and from all of the preceding – and all, except the first, has one, and only one, predecessor; and above all by the fact that between any two elements *the number of intermediaries is always finite.* A finite number of degrees separate them, and we go from one element to the other in a finite number of steps.

Finally, a very similar type is the scale of integers 'positive and negative', unlimited in both directions: a scale of the *type* ζ *(zeta);*

$$\ldots < 2 \text{ before} < 1 \text{ before} < \text{zero} < 1 \text{ after} < 2 \text{ after} < \ldots,$$

for which the characteristics are the same as for the ω type except that there is neither a greatest element nor a smallest.

The finite scales, and the ω and ζ scales are *discrete scales:* their common property is that there is always a finite number of intermediaries between two elements, this number being itself nul between an element and its successor. What distinguishes them is the *use of the superlative.* On the finite scales, there is a greatest element (a maximum) and a least element (a minimum), and every subset has a maximum and a minimum: the *absolute* superlative is always possible, but the *relative* superlative is not. A very large collection cannot always be ordered: 'still larger' does not always have a meaning. On the ω scale, a collection, however large, can be ordered. But if every subset has a minimum, certain of the subsets, and the whole scale, have no maximum: it does not always allow the absolute superlative. Finally, on the ζ scale, there is neither a maximum nor a minimum. We observe, furthermore, that on this scale the positioning of the cut between the numbers 'before' and the numbers 'after' is arbitrary.

The antinomy just pointed out as existing between the two superlatives is inescapable, in any case. It can be shown (we shall use it without proof) that a scale for which each subset possesses a maximum and a minimum is necessarily *finite.* No scale can thus have both the absolute superlative and the relative superlative for any of its subsets.

As opposed to discrete scales, we shall introduce *dense* (without gaps) scales, that is, scales such that between any two elements there is always at least one intermediary thus an *infinity of intermediaries,* since we can take the intermediaries of the intermediaries, *ad infinitum.* That the creation of dense scales was necessarily imposed on the intellect with respect to measurement scales is obvious. We can measure lengths, for example, only to within a centimetre. This means that we can compare lengths on a finite scale graded in centimetres. But we can subsequently increase the sensitivity of measurement if we have a scale graded in millimetres at our disposal. The accuracy of measurement will be further improved if we interdigitate gradations in tenths of millimetres, etc. . . . But however sensitive the comparisons that have made technical progress possible may be, the material scales of measure must always be finite. Nevertheless, it is useful, if not necessary, to have at our disposal a *vocabulary*, a *lexicon*, which permits us to designate, to *name*, all of the possible interdigitations, however fine they may be.

Among these vocabularies of 'infinitesimal smallness', one at least is familiar: the writing of numbers in decimals (considered here only from the point of view of order). Two 'words' written with the 12 sign alphabet composed of the 10 'letters', 0, 1, 2, 3, . . . 9, in their usual order, and the signs '.' and '–' can be compared at once on first reading. If I read '–4.212', I know that it precedes '1.314', which itself precedes '1.315'. It is sufficient to consider the alphabetic order (with special rules of '–', '.' and 0) of these words as we would a dictionary or a telephone book. Moreover, if I wish to insert a word between '1·314' and '1·315', I need only write '1·3141', for example.

The scale of decimal numbers is thus *dense* and *countable*. This latter term means that each of the elements can be designated by a word made up of a *finite* number of letters, taken from a *finite alphabet*. Similarly for the binary numbers: the conventions for writing them are the same as those of the decimal numbers, except that there are only two digits (letters) written as 0 and 1. The same is true of the scale of the *rational* numbers (irreducible fractions such as 12/5, 7/9, etc.); it is a countable scale. The alphabet here again has 12 signs; it is dense, since between any two fractions it is always possible to insert their half sums, for example.

The countable, dense scales, having neither maximum nor minimum constitute the *type* η (eta) of Cantor. In fact, he showed that between two such scales – that is, between two ordered lexicons that permit an infinite intercalation – it is always possible to define a bijection respecting the order of their elements, which was not at all obvious.

If the dense, enumerable scales give us a very good idea of what is generally meant by a 'continuum' in the social sciences, they are not, however, sufficient for locating all of the comparable magnitudes. They do permit us only to locate them with an ever increasing sensitivity through the use of *successive approximations*. For example, let us consider the series of approximations on the decimal scale of a number x, situated between 3 and 4:

«$3 < x < 4$» , «$3, 2 < x < 3, 3$» , «$3, 27 < x < 3, 28$» ,
«$3, 271 < x < 3, 272$» , etc.

We could continue indefinitely by writing longer and longer 'words', that is successive approximations. To what do these approximations converge? were we to follow such a process indefinitely? We certainly would not tend to an element of the decimal scale, since, by definition, we require a 'word' of infinite length, and the scale only includes words of finite length. The decimal scale, though often qualified as being 'without gaps' does present us with *discontinuities.* We can see at once that every series of approximations does not converge to an element of the scale, and such a series does no localize anything on this scale.

We are thus led to complete the dense and ennumerable scale by a set of 'words' of infinite length: a scale thus complemented is said to be *continous* (not to be confused with continuum), and its *type* is noted as λ (lambda), if it has neither a maximum nor a minimum. It is called θ (theta) if it does have a maximum and a minimum. The scale of points on a line segment, insofar as it is represented in Euclidean geometry, is of the type θ; the scale of all real numbers, rational and irrational, is of the type λ. For these scales we do not even have a lexicon that allows us to designate all of the elements. We can, in general, only name their approximations, because the elements being approached are themselves 'inexpressible' since the words designating them are of infinite length.

There are indeed other types of orders (see for example, Huntington, 1955), in the Bibliography), but with discrete orders, the dense, enumerable orders, and the continuous orders, we have essentially all the orders that have been actually used in the theory of measurements in the social sciences.

2. The additive calculus (abelian groups)

In the preceding paragraph we were reminded that there are many kinds of numbers:

1. the *integers,* whose set is written **Z** ($\mathbf{Z} = \{\ldots -2, -1, 0, 1, 2, \ldots\}$),
2. the *rationals* whose set is noted **Q**, ($\mathbf{Q} = \{p/q; p$ and q being integers, $q \neq 0\}$,
3. the *decimals,* whose set is noted **D**,
4. the *binaries,* whose set is noted **B**,
5. the *reals* (rational and irrational) whose set is noted **R**.

There are still others: for example, our system of writing numbers with ten letters leads to decimals; with two letters it leads to the binaries. But it is possible to apply the same procedure with an alphabet having any number (but at least 2) of letters. There are as many scales of numbers as there are integers; they are all of the same order type (type η).

These sets, **Z**, **Q**, **D**, **B**, and **R**, are not all of the same order type, and as we have seen the operations possible are not all the same. If in all of the sets we can always perform additions and subtractions (on the condition that we always consider positive and negative numbers) divisions are not always possible: the quotient of two numbers taken from **Q** or from **R** is always a number of the same set, which is not the case for **Z**, **D** or **B**. Similarly, in **R** we can always 'calculate' the square root of a positive number, but not in the other sets mentioned above.

What is common to all of these sets of numbers is that addition (and subtraction) is always possible. This is also true of multiplication, but this is a consequence of being able to perform addition. The rules of calculus by addition are:

1. The calculus is *associative*: $(x + y) + z = x + (y + z)$. The parentheses are unnecessary when the expression contains no more than three terms.
2. The number 0 is *neutral:* $x + 0 = 0 + x = x$.
3. Each number x has an *opposite:* $(-x)$: $x + (-x) = 0$.
4. The calculus is *commutative*: $x + y = y + x$; the order of the terms in an expression is irrelevant.

We have seen elsewhere (chapter 5: 'Monoids and groups') that the first three rules characterize the structure of a group. Moreover, the second and third of these rules (neutral and opposite) can be replaced by the condition: if a and b are given, there is a unique solution to the equation $a + x = b$; we write $x = b - a$, so that the subtraction can be performed. As there is an additional rule here, that of commutativity, we say that we have commutative groups, or (a synonym) abelian groups. In what is to follow, we shall simply use the term 'abelian'.

An abelian is thus a set A, with an associative, internal, binary operation, written +, for which there exists a neutral element, written 0; an opposite, written $-x$, for each element x, and which is commutative.

There are many examples of abelians, that is, sets of objects which can be combined among themselves according to a binary operation having the same rules given above for the addition of numbers. Let us consider one of the most familiar.

There are the positive rationals (strictly greater than zero) with respect to multiplication. This multiplication is certainly an associative and commutative operation, written × when the addition of the rationals is also being used. The neutral element is 1, and the 'opposite' of an element x is its 'inverse', $1/x$. We shall have occasion to return to this example (sections 4 and 5), but it should have already served to show that the difference between the 'additive' scale and the 'multiplicative' scale is not great; only the symbols and their sense are different; the rules of their calculus are the same.

The points of a 'pointed' plane or space constitute an abelian with respect to vectorial addition. We choose a point, the 'origin' 0, and to each pair of points A and B we make their 'sum' $A + B$ correspond by the parallelogram rule shown in Fig. 1. It is easy to show that the set of points in this space, with the previously defined addition, is indeed an abelian. The four rules of the abelians are satisfied. In particular, the origin point 0 is the neutral element, and the opposite $(-M)$ of a point M is its symmetric point with respect to the origin.

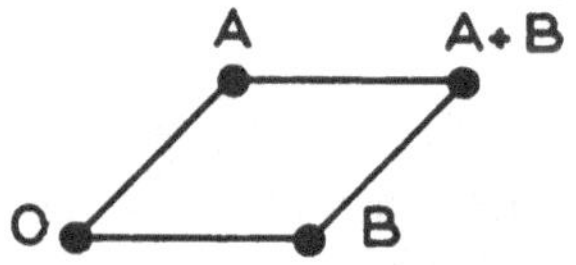

Fig. 1

The other example of an abelian is the set of number tables with the same number of rows and columns for which we will now define an 'addition'. Consider for example, the automobile production, for some country in numbers of vehicles constructed during five successive years, numbered: 1, 2, 3, 4, and 5. These data are usually given in table form:

Years	1	2	3	4	5
Production	x_1	x_2	x_3	x_4	x_5

In this table, x_1 stands for the number of vehicules produced in year number 1, x_2 in year 2, etc. All the 'information' contained in the table has been summarized by the *list* of successive productions: we can write this list as x where:

$$x = (x_1, x_2, x_3, x_4, x_5)$$

If we now consider the automobile production of a second country for the *same* years; we have a second list Y:

$$y = (y_1, y_2, y_3, y_4, y_5)$$

It is the usual practice in accountancy or in statistical economies to calculate the production for both countries together per year: it is a total which can be noted $X + Y$, for which the table can be obtained by adding the numbers entered in the column cells corresponding to the same year. This operation generates the list:

$$x+y = (x_1+y_1, x_2+y_2, x_3+y_3, x_4+y_4, x_5+y_5)$$

Similarly, the *difference* in the yearly production is often calculated: it is given by the list:

$$x-y = (x_1-y_1, x_2-y_2, \ldots, x_5-y_5).$$

The set of five integers in the column cells, with addition (and substraction) thus defined is an abelian. In particular, the neutral element is the list:

$$0 = (0, 0, 0, 0, 0)$$

and the opposite $-x$ of a table X is:

$$-x = (-x_1, -x_2, \ldots, -x_5).$$

Instead of integers, the x_i can be decimals or rationals, etc. This method of adding tables can be generalized to any number of columns and entries. In order that two tables of numbers be summable, it is necessary and sufficient that the number of entries be the same, and that the numbers written in the corresponding cells having the same index also have the *same significance.* We don't add cauliflowers and carrots, but the significances can be distinct for each of the cells having distinct indices: a number of dollars in the first, a number of kilowatt hours in the second, a number of inhabitants in the third, etc. It is a method of calculation very common in accountancy, in economies in descriptive statistics, etc. The reader can easily find from his own experience many examples where he has made similar calculations.

This example provides us with the opportunity of defining a very useful concept: the *direct sum* of several abelians. Being given two abelians A_1 and A_2, we consider the set of *pairs* $x = (x_1, x_2)$ where x_1 is any element of A_1, and x_2 is any element of A_2. We then define the addition of two pairs: $x = (x_1, x_2)$ and $y = (y_1, y_2)$ by the rule:

$$x+y = (x_1+y_1, x_2+y_2)$$

In the parentheses, the first addition is that of the abelian A_1 and the second is that of the abelian A_2. It is easy to show that the set of pairs, with this addition, is an abelian for which the neutral element is the pair $(0_1, 0_2)$, where 0_1 is the neutral of A_1 and 0_2 is the neutral of A_2. This abelian is the direct sum of A_1 and A_2, and is often written $A_1 \oplus A_2$.

The direct sum of three, four, etc. abelians is defined similarly. The set of the lists of five integers which we used for an earlier example is the direct sum of five abelians, all identical with the abelian (**Z**, +). (The notation (**Z**, +) represents the set **Z** of integers with its addition.)

There are still other examples of abelians. When we wish to count the hours numbered from 0 to 11 as they are pictured on the face of a clock or watch, we can add the time periods (in hours). In this addition we have: 3 + 2 = 5, but 7 + 6 = 1 because thirteen o'clock is read as 1 on the face of the watch similarly for 7 + 5 = 0 (12 = 0), and -3 = 9 (because 9 + 3 = 12 = 0). This is addition 'modulo 12' on the integers. We are again dealing with an abelian, but it is finite, having 12 elements: 0, 1, 2, 3, 4, 5, 6, 7, 8, 9, 10, 11; and we can set up the Pythagorean table of its addition. Moreover, this abelian is a cyclic group of order 12,[1] which can again be represented by the rotations of a turn by angles that are multiples of a twelfth around a point in the plane. The clock dial is a concrete example of this last representation.

The cyclic groups have been introduced in chapter 5, 'Monoids and groups'.

They are finite abelians, and for certain orders they correspond to very familiar modes of calculation, particularly with respect to time measures (nor is this by chance). Such are the order 2 (the half turns, the affirmation and negation in logic, etc.) the order 4 (the seasons), the order 7 (days of the week), the order 12 (hours and months), the order 24 (hours), and the order 60 (minutes and seconds).

However, there are finite abelians that are not cyclic; the smallest of these is the famous Klein group (see chapter 5).

If all of the *finite* abelians are not cyclic, all of them are, in fact, quasiabelian. A fundamental 'structure theorem' indeed asserts that *every finite abelian* (that is, those having only a finite number of elements) *is the direct sum of cyclic abelians.* For example, the Klein group is the direct sum of two cyclic groups of order 2 (see chapter 5).

In a finite abelian, as in any finite group, every element a is of finite order n: a added to itself successively n times equals 0 and generates the cyclic group of

1. The order of a group is the number of its elements.

order n. But an abelian can be infinite (have an infinite number of elements) and have elements of finite order, as is demonstrated by the example of the group of *all* the rotations of a plane around a point 0. The rotations of a rational fraction of a turn are of finite order. For example, the rotation of two-fifths of a rotation is of order 5; repeated five times consecutively it makes two complete rotations and returns to its initial position. It is equivalent to the nul rotation with the understanding that the rotations are counted 'modulo one rotation'. But the rotations of an angle, which is an irrational part of a complete turn, are of infinite order.

On the other hand, if we consider only those rotations for which the ratio to one rotation is rational, they constitute an abelian (the sum of two of these is a rational), which is a subgroup of the preceding. This abelian is *infinite*, but *all of its elements are of finite order;* such an abelian is sometimes said to be a *torsion* group, or to be periodic. These examples illustrate that one important principle of the classification of abelians is whether elements of finite order exist or not. An abelian is said to be torsion free if all of its elements (except 0) are of infinite order, as is the case for (**Z**, +), (**Q**, +), (**D**, +), the points on a plane as described on page 154, or the number lists on page 155. Each elements x, distinct from 0, added to itself an arbitrary number of times generates elements, each distinct, and consequently a group isomorphic to (**Z**, +) and composed of:

$$\{\ldots, -x-x, -x, 0, x, x+x, x+x+x, \ldots\}$$

which is convenient to write as:

$$\{\ldots, -2x, -x, 0, x, 2x, 3x, \ldots\}$$

On the other hand, the finite abelians and the rotational abelians are abelians with torsion: certain elements, possibly all, are of finite order.

But if to be with torsion or torsion free is one of the first classifications of the abelians, there is another classification which crosses the first, and it is directly related to the question of numeric scales. In any group G, we can define multiples of any x we choose, and write them according to the convention below:

$$x+x = 2x \qquad \underbrace{x+x+\ldots+x}_{n \text{ times}} = nx$$

$$-x = (-1)x \qquad -x-x = -2x, \quad \text{etc.}$$

By associativity we have for the + operation: $(n+m)x = nx + mx$. The addition of the second member is in G; that of the first member is performed according to the ordinary rules for the addition of the integers in (**Z**, +), if x is of infinite order. If x is of finite order k, then it is of 'modulo k' as we have seen in other examples.

If it is always possible to define multiples thus, such is not the case for division by an integer. By definition, an element x of an abelian A is divisible by a positive integer q if there exists an element y of this abelian such that:

$$qy = x \qquad \text{i.e.} \qquad \underbrace{y+y+\ldots+y}_{q \text{ times}} = x.$$

For example, in the Klein group (see its table, p. 134), no element except I is divisible by 2 because for every element x: $x + x = I$. But each element is divisible by 3, because for any x:

$$3x = x+x+x = I+x = x$$

In (Z, +), 12 is divisible by 12, 6, 4, 3, and 2; and in general, each number element of (**Z**, +) is divisible by its 'divisors' in the ordinary sense of the term. In the torsion free abelian (D, +) (the decimal numbers) each decimal number is equal to a fraction whose denominator is a power of 10. For example, $2.513 = 2513/10^3$. All of its elements are divisible by all the powers of 2, 5, and then 10; these powers are the only numbers that divide *all* of the elements of (**D**, +). Similarly, in the abelian of the binary numbers (**B**, +), the only divisors of all of the elements are the powers of 2.

On the other hand, certain abelians are such that every element y is divisible by any integer, as is the case for (**Q**, +) which is torsion free. Every rational p/q is divisible by any integer n, and the quotient is p/nq. Such is also the case for the group with torsion of the rotations around the point in a plane; a rotation of the angle θ is divisible by any integer n. The quotient is the rotation of the angle θ/n.

An abelian is said to be *divisible* if each of its elements is, as we have seen in the last two examples, *divisible by every positive integer.* Otherwise it is said to be not divisible. We have also seen that an abelian can be divisible with or without torsion, and non divisible with or without torsion.

The cross table below summarizes this typology. In each box we have indicated some, but not all, of the examples of the abelians satisfying the corresponding conditions:

	With torsion	Torsion free
Divisible	rotation groups	(**Q**, +) (**R**, +)
Non divisible	Finite abelians	(**Z**, +) (**D**, +)

The two classifications that we have just defined for the abelians: with and without torsion, divisible or not, are very far from constituting a complete typology, which in any case, remains to be completed. However, they are of major importance for our subject: the absence of torsion is, in effect, the necessary and sufficient condition for an abelian to be totally ordered (thus to constitute a scale). As to divisibility, it is related – when an abelian is ordered – to the *density* of the order.

3. Additive numerical scales (totally ordered abelians)

After our brief excursion into the diversity of order types and of the additive calculus types, it is now time to 'put the pieces together' since once the numerical scales have been given both a total order and an addition they are compatible with each other in the sense that to add a third element z to any other two elements x and y does not modify their order. Symbolically:

$$x < y \Rightarrow x+z < y+z, \qquad \text{for any } z.$$

In mathematics, the numerical scales are called *totally ordered abelians* that is, those abelians (additive) that have just been assigned an order compatible with addition in the sense that has just been defined.

A totally ordered abelian can obviously always be partitioned into two classes: the class P of elements superior or equal to 0 (0 included), or the *positive* elements, and the class N of those elements inferior to 0, the *negatives*. In fact, *the order being total,* every element is comparable to 0 and is either superior or inferior to it. The fact that order and addition are compatible has immediate consequences such as:

$$x \geqslant y \Rightarrow (x-y) \geqslant y+(-y) \qquad \text{i.e.} \quad x-y \geqslant 0$$

and conversely:

$$x - y \geqslant 0 \Rightarrow (x-y)+y \geqslant 0+y \qquad \text{i.e.} \quad x \geqslant y$$

In other words, x is superior to y if, and only if, the difference $x - y$ is positive. On the other hand, if an element is positive, its opposite is negative and conversely since:

$$x > 0 \Rightarrow x+(-x) > -x \qquad \text{i.e.} \quad 0 > -x$$

The abelians with which we are already familiar, the integers (**Z**, +), the decimals (**D**, +), the binaries (**B**, +), the rationals (**Q**, +), the (strictly) positive rationals with respect to multiplication (**Q***, ×), and the reals (**R**, +), all ordered in the usual way, are totally ordered abelians.

We note that all are *torsion free*. Indeed, every ordered abelian is necessarily *torsion free*, for, if x is any element and if, for example, $x > 0$, this inequality implies:

$$x+x > x+0 \qquad \text{i.e.} \quad 2x > x,$$

This itself implies:

$$2x+x > x+x \qquad \text{i.e.} \quad 3x > 2\,x, \text{ etc.}$$

Thus, the succession of multiples of a positive element x constitutes an order of type ζ:

$$0 < x < 2x < 3x < \ldots < nx < \ldots$$

As we have seen, $-x$ is thus negative. An analogous argument shows finally that the multiples of x and $-x$ are all distinct and constitute an order:

$$\ldots < -2x < -x < 0 < x < 2x < 3x < \ldots$$

We can then ask under what conditions can an additive calculus be given an order (total or even partial) compatible with itself, and consequently, be made into a numerical scale. Which abelians are 'orderable'?

It is clear from the preceding that for an abelian (not ordered) to be orderable it is necessary that it be torsion free. We can show (and will use it here without proof) that this condition is also sufficient: any torsion-free abelian can be assigned an order compatible with addition. The reasoning that proves that any ordered abelian is torsion free also shows that any ordered abelian contains at least one *subgroup isomorphic to the group (Z, +)* of integers, which is also, so to speak, the 'smallest' of these abelians (the smallest numerical scale).

But we also note that among the examples of numerical scales only one is of the discrete type with respect to order, (**Z**, +), and only one is of the continuous type, (**R**, +), while several are of the dense and countable type: (**Q**, +), (**D**, +) and (**B**, +).

That we have encountered but one example of numerical scales of the discrete type is not by chance; there are no others. The uniqueness (to within an isomorphism) of the totally ordered abelians of the discrete type is easily demonstrated. Thus, let $(A, +, \geqslant)$ be an abelian with a discrete order. It is necessarily of the type ζ because its type can be neither finite (it is ordered and thus torsion free), nor of the type ω since to each element $x > 0$ corresponds $-x < 0$; it cannot therefore have a minimum in A. Furthermore, let us designate in the order of A by $s(x)$, the unique successor of each element x, and by $p(x)$ its unique predecessor. Thus, with respect to 0, the neutral element of addition, the elements of A are in the order:

$$\ldots < p(p(0)) < p(0) < 0 < s(0) < s(s(0)) < \ldots$$

Whereas, for each x we have:

$$s(x) = x + s(0)$$

And indeed:

$$s(0) > 0$$

Thus, the order being compatible with addition:

$$s(0)+x > 0+x = x \qquad \text{i.e} \quad s(0)+x > x$$

And, as $s(x)$ is the smallest element superior to x:

$$s(0)+x \geqslant s(x)$$

But suppose that there were an intermediate y between $s(x)$ and $s(0)+x$:

$$s(0)+x > y > s(x)$$

Then we would have:

$$s(0) > y-x > s(x)-x > 0$$

Thus, $y - x$ would be intermediate between 0 and its successor $s(0)$, which contradicts the definition of successor. By setting $s(0) = 1$, we see that $s(x) = x + 1$.

Consequently, the order of the elements 0, $s(0)$, $s(s(0))$, etc. is expressed in *terms of addition* by:

$$0, 1, 1+1, (1+1)+1, \ldots$$

and these are the positive integers. An equally simple argument shows that

$$p(0) = -1$$

since

$$0 = s(p(0)) = p(0)+1$$

which completes the proof.

The multiplicity of dense, countable orders can be easily appreciated if we see that every ordered group that is divisible by a given integer is, for whatever the value of that integer, assigned a dense order. Suppose that a totally ordered abelian A is, for example, divisible by 2: for every element x of A there exists an element y such that $y + y = 2y = x$.

First of all, this element is *unique,* for if z is also such that $z + z = x$, and if we have $z < y$, for example (the order being total, z is comparable to y), then:

$$x = z+z < z+y < y+y = x \qquad \text{i.e.} \quad x < x \text{(strictly)}$$

This is contradictory. The unique element y can be written $x/2$.

Then, if x is positive we have:

$$0 < x/2 < x$$

Indeed, $y = x/2$ cannot be negative (since the sum of two negative elements is negative), neither is it nul, or we would have:

$$x = y+y = 0+0 = 0;$$

thus contradicting $x > 0$.

Finally, if $y \geqslant x$ then $x = y + y \geqslant x + y$. But since $y > 0$, $x + y > x + 0 = x$, leading again to the contradiction that $x > x$. Similarly, when x is negative we have: $x < x/2 < 0$.

It follows that we can always insert an element between 0 and any element x of A: $x/2$, in fact. But if a and b are any two elements of A, and if $b > a$, for example, we then have $b - a > 0$. Between $b - a$ and 0 we can insert $(b - a)/2$

$$b-a > (b-a)/2 > 0.$$

But then, $a + (b - a)/2$ is between a and b, since:

$$b = (b-a)+a > (b-a)/(2)+a > 0+a = a$$

This proves the density of the order of A. Furthermore, each element x of A can be divided by any power of 2. Indeed, $x/2$ can in turn allow an element t such that: $t + t = x/2$. But according to the associativity of addition:

$$t+t+t+t = x/2+x/2 = x$$

Therefore $4t = x$, and x is divisible by 4. We can set $t = x/4 = x/2^2$. But then $x/2^2$ is in turn divisible by 2, etc. . . . Thus to each element x of A corresponds a succession of quotients by the powers of 2: $x/2$, $x/2^2$, . . . $x/2^n$, . . . And an analogous argument can show that if $x > 0$, this series is decreasing:

$$0 < \ldots < x/2^n < \ldots < x/2^3 < x/2^2 < x/2 < x.$$

Thus it follows that any ordered abelian divisible by 2 contains a subgroup isomorphic to that of the binary numbers, since it contains with each element x all of the elements of the form 'nx plus a sum of terms of the form $x/2^i$.' (**B**, +) is thus 'the smallest' of the totally ordered abelians which are divisible by 2.

The demonstration of the uniqueness, to within an isomorphism, of the totally ordered abelians having an order of the *continuous* type requires more mathematical technique, and we will omit it here. But this uniqueness should be borne in mind since it has some important consequences for us in section 4.

Observing that in the concept of *numeric scale* there was that of addition, hence an abelian group, and that of comparison, hence that of total order, we have been led to study the totally ordered abelian groups: every numeric scale is one. But the converse is not true; the concept of totally ordered abelians *does not exhaust our usual idea of a numerical scale.* We need a least one additional condition, as our next example illustrates.

Let us consider the direct sum **Z** + **Z**, the abelian being assigned its natural order, and to simplify the exposé, let us represent the pairs of integers (x_1, x_2) by the points having integer coordinates on a plane of two axes. Let us give **Z** + **Z** the total order defined by:

$$(x_1, x_2) \geqslant (y_1, y_2) \Rightarrow (x_1 > y_1 \text{ or } (x_1 = y_1 \text{ et } x_2 \geqslant y_2))$$

In other words, to compare (x_1, x_2) and (y_1, y_2), we first compare their first components and arrange them in the order of these components. If these components are equal, we put the pairs in the order of their second components. It is easy to verify that this relation does indeed have the desired properties for an order. (In fact, it is one of the lexicographic orders often referred to in section 1; here it is the order of 'words' of two letters with the set of integers as an 'alphabet'.[2]) Moreover, this order, by definition total, is indeed compatible with the addition of the pairs:

$$[(x_1, x_2) \geqslant (y_1, y_2)] \Rightarrow [(x_1+z_1, x_2+z_2) \geqslant (y_1+z_1, y_2+z_2)]$$

for any pair (z_1, z_2).

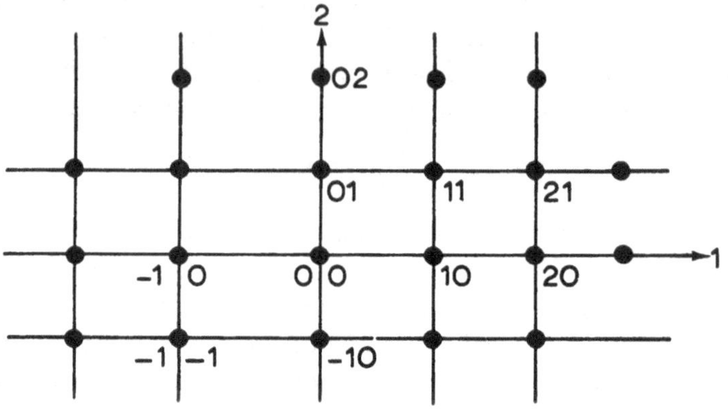

Fig. 2

On the lattice of Fig. 2, the points of the plane are thus ordered in such a fashion that all those on the same vertical succeed each other according to an order of type ζ, but come before those of the vertical which lies immediately to the right. For example, the point (0,1) precedes its multiples:

$$2\,(0, 1) = (0, 2); \qquad 3\,(0, 1) = (0, 3) \text{ etc.}$$

but all of these multiples are themselves below the point (1,0).

This totally ordered abelian, which hardly resembles the usual numerical scales, is such that for certain pairs of its elements, let us say a and b, where $0 < a < b$, for example, all of the multiples na of a are less than b. On the numerical scales, on the other hand, the succession of multiples of an element is not only unlimited, but unbounded as well; there are multiples greater than any given number b, however large. Symbolically:

If $0 < a < b$, there is at least one positive integer n such that $na > b$.
Or, if for any positive integer n $0 \leqslant na \leqslant b$, then $a = 0$.

2. This is also the order that we adopt in the practical problems of choice in multiple criteria when these criteria are placed in a hierarchy. The example given here thus corresponds to a manner of ordering which is found in many applications: lexicons, theory of choice, etc.

These scales thus satisfy the condition called Archimedean; they are the abelians, totally ordered and Archimedean. It is remarkable that now all of the conditions have been given: that is, totally ordered, Archimedean abelians are necessarily subgroups of the additive group (**R**, +) of the real numbers. This is true even of every totally ordered Archimedean group (such a group is consequently commutative). We can thus consider that the equivalent mathematics of numeric measurement scales is constituted by the class of totally ordered groups for which the order is Archimedean. The abelian group of real numbers contains all of these groups as subgroups. They are Archimedean subgroups. Among them are those we know well: (**Z**, +), (**D**, +), (**B**, +) (and those which can be constructed is a fashion analogous to (**B**, +)), (**Q**, +) and (**R**, +) itself, the most fundamental. In a sense, they are the materials from which all the others are constructed (by direct sum, in particular).

4. The additive and multiplicative scales

One important question is to know all of the ways of transforming a number scale into another; by transform we mean the way in which the order and the addition of the first scale map on the order and the addition in the second scale. The interesting transformations are thus the homomorphisms of one scale to another.

In light of what has just been said, we will have such transformations if we know how to map the ordered abelians (**R**, +, $\geqslant$) of reals into itself by means of a mapping f: $\mathbf{R} \to \mathbf{R}$, having the properties of monotony (conservation of order) and of linearity (conservation of addition):

Monotony: $x > y \Leftrightarrow fx > fy$,
Linearity: $f(x+y) = fx + fy$.

The construction of all of these homomorphisms of (**R**, +, $\geqslant$) into itself, or endomorphisms of (**R**, +, $\geqslant$), is easy. We first observe that the image of 0 is:

$$f(0) = 0$$

because

$$f(x) = f(x+0) = f(x)+f(0)$$

Let us choose for the image of 1 an arbitrary positive real because of the monotony:

$$1 > 0 \Rightarrow a = f(1) > f(0) = 0$$

Then we have:

$$f(2) = f(1+1) = f(1)+f(1) = a+a = 2.a$$

Generalizing this calculus, we have for every positive integer:

$$f(n) = a+a+\ldots+a = na$$

And since:

$$0 = f(0) = f(1)+f(-1) = f(1+(-1)) = a+f(-1)$$
$$f(-1) = -a$$

and for every negative:

$$f(-n) = -na$$

We have thus the images of all of the integers. Now for the rational $1/2$, its image $f(1/2)$ is such that:

$$a = f(1) = f(1/2+1/2) = f(1/2)+f(1/2)$$

thus that $f(1/2) = a/2$

and in general:

$$f(1/n) = a/n.$$

Finally, for any rational $p/q = \underbrace{(1/q + 1/q + \ldots + 1/q)}_{p \text{ times}}$:

$$f(p/q) = p(a/q) = (p/q)a$$

We thus have the images of all of the rationals. Now let x be irrational. Let us suppose that its image $f(x)$ is different from xa, greater, for example. As there are approximations as close as we wish to x by the rationals, we can procede in such a way that for some rational r, greater than x, $f(r) = ra$ will lie between xa and $f(x)$. We have then that $r > x$ and $f(r) < f(x)$ which violates the monotony condition. Moreover, the only *endomorphisms* of the abelian $(\mathbf{R}, +, \geqslant)$ are the mappings of the form: $x \mapsto xa$ (where $a > 0$). Each of these endomorphisms are entirely determined by the *image a of 1.* The fact that this image is arbitrary signifies that the unit is equally arbitrary on a numerical additive scale.

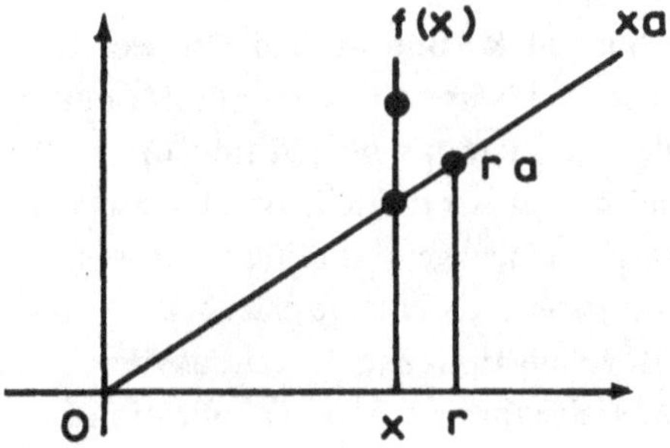

Fig. 3

There are four comments on this result:

1. To take the coefficient of a negative does not modify much; this simply amounts inverting the sense of the order.

2. The image of any x is the product xa in the *multiplication* of x by a. This product is possible because, as we all know, the numbers can be multiplied among themselves. But we could have supposed that the multiplication of the real numbers were *not defined a priori*, and written as a *definition* of multiplication that the product of two reals x and y, 'xy', is the image of x in the endomorphism of (R, +, ⩾), for which the image of 1 is y.
On the basis of this definition we can find all of the properties of multiplication.

3. The preceding constructions show that if we look for the endomorphisms of (**Z**, +), (**D**, +) or (**Q**, +) it suffices to suppose that they are endomorphisms for *the only addition* in order that they assume the form: $x \mapsto xa$. The hypotheses of monotony does not become fundamental until we wish to determine the images of the irrationals.

4. The transformation just found is *bijective* in (**R**, +, ⩾), and even in (**Q**, +, ⩾). If $y = xa$ is the image of x, then $x = y/a$ is the image of y in the inverse transformation. But this is true neither in **Z** nor in **D** nor in the analogous abelians. If, for example, in **Z** we choose a as the image of the integer 5, the image of **Z** is composed only of the multiples $5x$ of the integer 5. This is only a proper subset of **Z** and not the whole of **Z**.

We mentioned earlier that with respect to the only *multiplication*, the set **R*** of the strictly positive rationals is an abelian, and it is even an ordered abelian since:

$$x < y \Leftrightarrow xz < yz \qquad \text{(since } z > 0\text{)}$$

This is also the case for the set **R*** of *positive* reals. With respect to multiplication and the usual order, it is an ordered abelian (**R***, +, ⩾), whose order is continuous as is the whole of **R**. But we had also noted that all the totally ordered continuous abelians are isomorphic. From this point of view, to describe a continuous numerical scale in terms of addition or multiplication is only a question of language and not of 'structure', for they are strictly the same. The additive and multiplicative language(s) here are absolutely equivalent with respect to the rules of comparison and of the calculus.

This is an important acquisition and it can obviate numerous sterile discussions on the nature of scales that we find in applications.

However, it is also necessary to know the good 'dictionaries' which allow us to go from one language to another. That is, the functions f of the additive abelian $(\mathbf{R}, +, \geqslant)$ in the multiplicative abelian $(\mathbf{R}^*, x, \geqslant)$ that transform the order of the former into that of the latter, and addition into multiplication:

$$x > y \Rightarrow fx > fy$$
$$f(x+y) = f(x) \times f(y)$$

These functions, called *exponential functions,* are constructed *mutatis mutandis* as the supra linear functions. This time, the image of 0, the neutral element of the first abelian is 1; the neutral of the second: $f(0) = 1$. The image of 1 is an arbitrary positive real b, but superior to 1 (monotony):

$$f(1) = b > 1 = f(0)$$

The image of 2 is then:

$$f(2) = f(1+1) = f(1) \times f(1) = b \times b = b^2$$

And in general

$$f(n) = b^n = \underbrace{b \times b \times \ldots \times b}_{n \text{ times}}$$

Similarly:

$$f(-n) = 1/b^n.$$

(By analogy with the notation b^n we can write:

$$1/b^n = b^{-n} \quad (f(-n))$$

We also note that in this notation: $b^0 = f(0) = 1$. For the rationals: $f(1/2) = \sqrt{b}$ (square root of b); in exponential notation becomes:

$$f(1/2) = b^{1/2}$$

and more generally, for a rational p/q, and for any real x, we can write:

$$f(p/q) = b^{p/q}, \qquad f(x) = b^x$$

Here again each function is determined by the only image b of 1; b is called the *base* of the exponential. We should note further that in the exponential notation, the fundamental relation defining the exponential:

$$f(x+y) = f(x)\times f(y)$$

is written:

$$b^{x+y} = b^x \times b^y$$

which is the first rule of the *exponential calculus.*

The 'dictionaries' which transform the additive language to the multiplicative language are thus exponential functions. Conversely, we construct the dictionaries from the multiplicative language, to the additive language; they are the inverse functions of the exponentials, the logarithmic functions. By definition, the logarithm to the base b of a positive real x is the real, positive or negative y, for which x is the exponential to the base b, written:

$$\log_b x = y \Leftrightarrow b^y = x$$

According to the definition of the logarithmic functions we have the usual rules of logarithmic calculus such as:

$$\log_b 1 = 0 \quad (b^0 = 1), \ \log_b 1/x = -\log_b x \quad (b^{-y} = 1/b^y)$$

and above all we have:

$$\log_b (x\times y) = \log_b x + \log_b y$$

which expresses the definition: the transformation of products into sums.

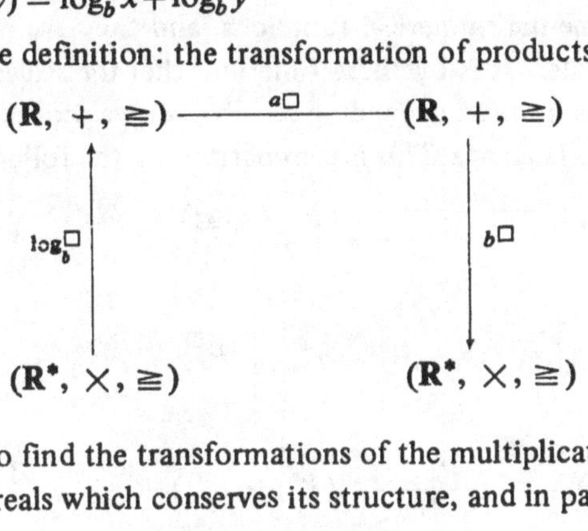

Fig. 4

It remains for us to find the transformations of the multiplicative abelian ($\mathbf{R}^*$, x, $\geqslant$) in the positive reals which conserves its structure, and in particular transforms multiplication into multiplication. This transformation can be effected through the use of addition as indicated in the accompanying diagram (Fig. 4). Each of these three successive transformations preserves the order, and their resultant as well. Moreover, multiplication is transformed into addition by the logarithm. Addition in turn remains addition by the linear function, and is finally transformed into multiplication by the exponential. What is the form of this composite transformation? A positive real x is first transformed into $\log_b x = y$:

$$x \mapsto \log_b x = y$$

y is in turn transformed into $ya = z$:

$$y \mapsto ya = z = ay = a \log_b x = \log_b x^a$$

Finally, z is transformed into b^z:

$$z \mapsto b^z = b^{(\log_b x^a)}$$

But according to the definition of logarithim to the base b:

$$b^{(\log_b \square)} = \square$$

Finally, we have $z = x^a$ and the transformation sought is a power function:

$$x \mapsto x^a$$

According to the definition of this function, we have in particular for every pair of real positives x and y:

$$(x \times y)^a = x^a \times y^a.$$

The four functions: linear (xa), exponential (b^x), logarithim ($\log_b x$), and power (x^a) are the fundamental numerical functions, and they are intrinsically related to the numerical scales. It is by these functions that the scales remain invariant, or correspond to each other, depending on whether we use the additive language or the multiplicative language. This is summarized by the following diagrams:

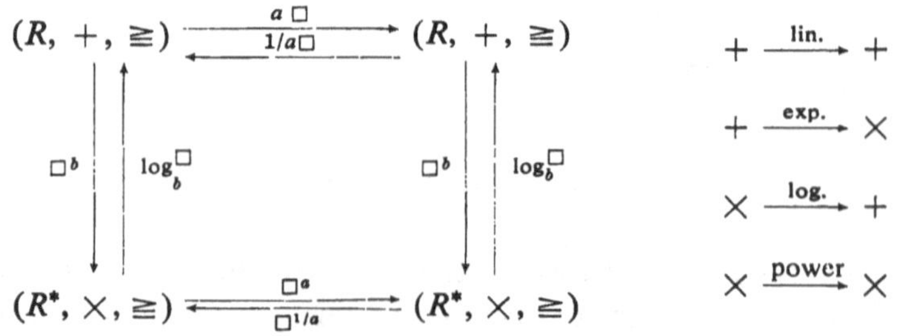

Fig. 5

5. Interval scales. Affine scales

'Numerical scale' is synonymous for us with abelian, totally ordered and Archimedean scale. But it sometimes happens in applications that a numerical scale is not presented directly in this form, and that we are confronted with another presentation of ordered abelians.

One of the most usual presentations in the social sciences is that which we sometimes call 'interval scales'. We compare objects (sensations, for example) between themselves, and we have first a total order, noted here as $\geqslant$, of comparison. But on the other hand, we establish a relation between pairs which can be written:

$$a : b \succ c : d$$

which is to be read:

	a is to *b* more than *c* is to *d*
or:	*a* differs from *b* more than *c* differs from *d*
or:	*a* surpasses *b* more than *c* surpasses *d*

etc.

The relation is assumed to be a total quasiordering (see chapter 2: 'Ordering and classification') on the pairs; this quasiordering being compatible with the order of comparison $\geqslant$ between objects in the sence that:

$$a < x < y \Rightarrow y : a \succ x : a$$

In this quasiordering, we have equivalence classes and we can write:

$$a : b \equiv c : d$$

the equivalence of the pairs (a, b) and (c, d) which is read: '*a* is to *b* as *c* is to *d*', and we can call this an *analogy*.

We then make certain hypotheses concerning this equivalence. Here, for example, is a set of possible hypotheses:

1. $a : b \equiv c : x$ determines a unique element x if a, b and c are given.
2. $a : b \equiv c : d \Leftrightarrow a : c \equiv b : d$ (the so called 'bisymmetric law').
3. $a : b \equiv c : d \Leftrightarrow d : b = c : a$ (exchange of extremes).

Hypotheses (1) signifies (taking into account hypotheses 2 and 3) that in an analogy, three given terms determine the fourth. It has in particular the consequence that : $a : a \equiv b : c \Rightarrow b = c$. Indeed, according to (2), this analogy is the same as: $a : b \equiv a : c$; c is thus the unique solution of; $a : b \equiv a : x$, and this solution is obviously b since the relation written '$\equiv$' is an equivalence (it is reflexive).

As for hypotheses (2) and (3), they can be formulated by saying that the analogy is invariant by the transformations of the Klein four-group, which we saw in an earlier table (chapter 5: 'Monoids and groups'). Let us now consider the following transformation M on a quadruplet *abcd* (bisymmetric law):

$$abcd \xrightarrow{M} acbd$$

and then of the transformation, E (exchange of extremes):

$$abcd \xrightarrow{E} dbca$$

It is easy to see that:

$$E \text{ followed by } E = M \text{ followed by } M = I$$

where I means 'change nothing' (identical transformation). Similarly:

$$M \text{ followed by } E = E \text{ followed by } M$$

Let us call this last transformation S; its effect on the quadruplet is to reverse the order:

$$abcd \xrightarrow{S} dcba$$

Thus, the set $\{I, E, M, S\}$ with its operation of composition (a transformation followed by another) is the Klein four-group as the reader can verify without difficulty by constructing the table.

Using this system of hypotheses, if we choose once and for all an arbitrary element 0, and if we choose any two elements a and b of the scale, the analogy: '$0 : a \equiv b : x$' determines a unique element x. We can write: $x = a \oplus b$, and we can then easily show that with respect to the operation $\oplus$ so defined, the scale is a *totally ordered abelian group,* for which the element 0 is the neutral element, and by means of the subtraction in this abelian, the analogy can be written:

$$a : b \equiv c : d \Leftrightarrow a - b = c - d$$

Just to give an idea of the proof, let us demonstrate, for example the commutativity.

Let $x = a \oplus b$, and $y = b \oplus a$, we have thus by definition:

$$0 : a \equiv b : x \quad \text{and} \quad 0 : b \equiv a : y$$

But according to the bisymmetric law, the first of these analogies is equivalent to:

$$0 : b \equiv a : x$$

Thus '≡' begin an equivalence (transitive):

$$a : x \equiv a : y.$$

which as we have seen implies: $x = y$.

The proof of the other properties of ⊕ (associativity, existence of a neutral; 0, and an inverse) is analogous and is left to the reader.

We have spoken here of interval scales (scales constructed by the comparison of intervals) because we had chosen to speak the 'additive' language, If we choose to speak in 'multiplicative' we have the same axiomatics, or axiomatics which lead to the same results for the analogies and it will sometimes be called the axiomatic of a ratio scale. Again, the difference is in the words, not the structure. What is essential is the result: using proper hypotheses, a given interval ratio scale is the same as that of an additive (or multiplicative) numerical scale.

We also noted that a totally ordered abelian which is *equivalent* to an interval scale satisfying an appropriate axiomatic is only determined by the choice of *the neutral element* 0. The origin of the equivalent additive scale with respect to an interval scale is arbitrary (or 'multiplicative': the unity of a multiplicative scale is arbitrary). The transformations leaving an interval scale invariant are not the same as those which leave the additive numerical scales invariant (that, is the linear functions). What then are these transformations?

We must determine all the mappings f of an abelian that is totally ordered into itself and which:

1. conserves the order: $x > y \Rightarrow f(x) > f(y)$; and
2. conserves the analogies (equality of intervals):

$$x - y = z - t \Rightarrow f(x) - f(y) = f(z) - f(t)$$

For condition 2, if we take the neutral element 0 of the abelian as the fourth term, the above expression can be written:

2′) $\qquad x - y = z \Rightarrow f(x) - f(y) = f(z) - f(0).$

If we now write for every x: $g(x) = f(x) - f(0)$, the function g, which differs from f by the constant $f(0)$, is monotonic. Making use of this function 2′ can be written:

2″) $\qquad x - y = z \Rightarrow g(x) - g(y) = g(z)$

that is,

$$x = y+z \Rightarrow g(x) = g(y)+g(z).$$

Hence: $g(y + z) = g(y) + g(z)$. g is thus a *linear function*, and we therefore have:

$$g(x) = ax$$

and setting $f(0) = b$, we finally obtain:

$$f(x) = ax+b.$$

Those functions, represented by straight lines in cartesian graphs, are called *affine functions*, just as we call *affine space* those structures defined by the equality of the differences on an abelian. The 'interval scales' are, in the last analysis, *affine numerical scales*. They are defined up to the choice of the origin and of the unity of measure.

I should like to say a few words abouth another presentation which also is tending to become classic. The idea is that of the 'mean' of two elements of a totally ordered set, a scale. Given two elements x and y, we can compare them, and we have, for example $x < y$. We then ask ourselves which is the element on the scale that can represent the 'mean', or the 'central value' or the 'median' of x and y. We require that to every pair of elements x and y there will correspond a 'mean' situated between them,

$$x < x \cdot y < y.$$

Consequently, it is clear that a scale permitting an operation of 'mean' must be dense. As for the axiomatic properties demanded by such an operation, we will look for them among the classical properties such as the usual operations of mean, 'the weighted mean'. For example:

$$x \cdot y = px+(1-p)y \qquad (0 < p < 1)$$

A possible set of axioms is, for example:

1. Compatibility with the order: $x < y \Rightarrow \begin{cases} x \cdot z < y \cdot z \\ \text{and} \\ z \cdot x < z \cdot y \end{cases}$ for all z
2. Idempotence: $x \cdot x = x$
3. Bisymmetric law: $(x \cdot y) \cdot (z \cdot t) = (x \cdot z) \cdot (y \cdot t)$
4. In the equality $x.y = z$, two given terms determine the third.

1 and 2 imply, in any case, that $x.y$ is an intermediary of x and y. The result is that a scale permitting an operation of mean satisfying these conditions can be constructed as a totally ordered abelian (here too, the choice of 0 is arbitrary, and we are dealing with an affine space). Furthermore, by using the operation '+' of this abelian we have:

$$x\cdot y = f(x)+g(y)$$

where f and g are two automorphisms of the abelian such that for every x:

$$f(x) + g(x) = x.$$

Finally, if we assume the condition of Archimedes, the only endomorphisms for the Archimedean abelians are linear functions; we have thus:

$$x \mapsto ax,$$

and we, find the usual 'weighted means'.

Other presentations are possible, such as the one based on the axiomatic of the *ternary 'betweeness' relation*: 'x is between y and z' where the notion of an interval is also in use. The principal interest in applying these variants is to furnish the practitioner with a whole collection of criteria permitting him to recognize, in a particular situation where he thinks he is able to construct a set of *observable* phenomena in terms of scales, whether it can be done and which scale should be used.

The questions which should be asked are in fact the following: first (naturally) whether these phenomena can be *compared* with each other, and whether the relation of comparison really has the desired properties: that is transitivity, totality (since, in general the *ex-aequo* of equivalences are not on the scale but among the *observations*, the question of symmetry or of antisymmetry is not relevant). If the answer is yes, the practitioner is dealing with a total quasi-ordering for which the classes are totally ordered (constituting a scale).

Secondly, he must then ask himself: what is the type of this order? In particular, is it discrete or dense?

Thirdly, he must determine whether there is a significant way of combining these phenomena with each other in such a fashion that a well determined third corresponds to any two given ones. Then he must look for *combinatoric properties* or algebraic properties of this combination from among those which lead to numeric structures: associative, or not: commutative, or not: bisymmetric, or not, etc. If the practitioner is capable of answering all of these questions, the construction of the corresponding scales and knowing whether it is or is not a usual numeric scale, or some other object, is but a routine mathematical affair.

In this article I have indicated what the response to these questions is in some cases; but in many others they are probably known or easy to determine. The difficulty lies in having a theory and knowledge of the observed facts so that the mathematician can be furnished with the 'prerequisite conditions', but this is not a mathematical problem.

Bibliography

1. Mathematics

1.1 on total ordering (section 1):

Barbut, M., and Monjardet, B., *Ordre et classification: algèbre et combinatoire.* Paris. Hachette, 1970.

1.2 Finite orderings:

Frey, L., *Techniques ordinales en analyse de données: algèbre et combinatoire.* Paris, Hachette, 1972.

Ordres totaux finis. Paris–la Haye, Mouton/Gauthiers-Villars, 1970.

1.3 Infinite orderings:

Cantor, G., *Contributions to the founding of transfinite numbers.* Dover, 1955.

Hungtington, E. V., *The continuum and other types of serial orders.* Dover, 1955.

1.4 On abelians (section 2):

1.5 Finite abelians:

Carmichael, R. D., *Groups of finite order.* Dover, 1956.

Ledermann, W., *Introduction to the theory of groups of finite order.* London, Oliver and Boyd, 1953.

Marshall Hall, Jr., *Theory of groups.* New York, MacMillan, 1959.

1.6 Infinite abelians:

Fuchs, L., *Abelian groups.* New York, Pergamon Press, 1967.
Kaplansky, I., *Infinite abelian groups.* Michigan, The University of Michigan Press, 1965.

1.7 On ordered abelians (section 3):

Fuchs, L., *Partially ordered algebraic systems.* New York, Pergamon Press, 1963.
Kaplansky, L., *Infinite abelian groups.* Michigan, The University of Michigan Press, 1965.

1.8 On numerical functions (section 4):

Guilbaud, G. Th., *Mathématiques.* Paris, P.U.F., 1966.

1.9 On affine numeric scales (section 5):

Papy, G., *Géométrie affine plane et nombres réels.* Paris, Gauthier-Villars, 1962.

1.10 More technical works:

Artin, A., *Algèbre géométrique.* Paris, Gauthier-Villars, 1962, chap. 1 and 2; English ed: *Geometric algebra (interscience), 1967.*
Frink, O., 'Symmetric and self-distributive systems', *A.M.M.*, 1955, vol. 62, pp. 697–707.
Fuchs, L., *Partially ordered algebraic systems.* New York, Pergamon Press, 1963.
Luce, R. C., 'Algebraic systems of measurement', in *Algebraic models in Psychology.* O.T.A.N., 1969, pp. 265–298.
Sheppard, J. A. H., 'Betweeness groups', *J. London Math. Soc.*, 1957, vol. 32, pp. 277–285.

2. Scales in social science methods

Barbut, M., d'Adhemar, C., Leclerc, B., and Jullien, P., *Mathématiques élémentaires; applications à la statistique et aux sciences sociales.* Paris, P.U.F., 1973.
Boudon, R., and Lazarsfeld, P., *Le vocabulaire des sciences sociales.* Paris-La Haye, Mouton, 1965.

Coombs, C. H., *A theory of data.* New York, John Wiley.

Daval, R., *Traité de pscyhologie sociale.* Paris, P.U.F., 1964.

Degenne, A., *Techniques ordinales en analyse des données; statistique.* Paris, Hachette, 1972.

Edwards, A. L., *Technics of attitude scale construction.* New York, Appleton.

Matalon, B., *L'analyse hiérarchique.* Paris-La Haye, Mouton/Gauthier-Villars, 1965.

Reuchlin, M., *Les méthodes quantitatives en psychologie.* Paris, P.U.F., 1962.

subject index

methods and models

in the social sciences

1. **A Theory and Procedure of Scale Analysis**
   With Applications in Political Research
   by R. J. Mokken
   1971, XII + 354 pages. Clothbound
   ISBN: 90-279-6882-9

2. **Foundations of Theory-Formation in Criminology**
   A Methodological Analysis
   by C. I. Dessaur
   1971, XIV + 152 pages. Clothbound
   ISBN: 90-279-6887-X

3. **Mathematics of Decision Theory**
   by Peter C. Fishburn
   1972, 104 pages. Paperbound
   ISBN: 90-279-7142-0

4. **Initiation to the Mathematics of the Process of Diffusion, Contagion and Propagation**
   by J. P. Monin, R. Benayoun and B. Sert
   1976, VI + 94 pages. Paperbound
   ISBN: 90-279-7611-2

5. **Combinatorics, Graphs and Algebra**
   by Centre de Mathématique Sociale,
   École des Hautes Études en Sciences Sociales
   1976, 182 pages. Paperbound
   ISBN: 90-279-7511-6

6. **Multidimensional Preference Scaling**
   by Gordon G. Bechtel
   1976, X + 162 pages. Paperbound
   ISBN: 90-279-7592-2

mouton · the hague · paris

FSC
www.fsc.org
MIX
Papier | Fördert gute Waldnutzung
FSC® C083411

Kolibri 360 GmbH
Ferdinand-Jühlke-Straße 7
99095 Erfurt, Deutschland
produktsicherheit@kolibri360.de